The Amateur Astronomer

PATRICK MOORE

W · W · Norton & Company
New York   London

Printed in the United States of America.

Manufacturing by AG: Halliday.

Eleventh Edition

Library of Congress Cataloging-in-Publication Data

Moore, Patrick.
  The amateur astronomer / Patrick Moore.—11th ed.
    p.   cm.
    1. Astronomy.   2. Astronomy—Observers' manuals.   I. Title.
  QB45.M728   1990
  520—dc20   89–35278

ISBN 0-393-02864-X

W. W. Norton & Company, Inc., 500 Fifth Avenue, New York, N.Y. 10110
W. W. Norton & Company Ltd., 37 Great Russell Street, London WC1B 3NU

1 2 3 4 5 6 7 8 9 0

CONTENTS

The Amateur Astronomer

INTRODUCTION

When I wrote the first edition of this book, in 1957, the Space Age had not begun. Even artificial satellites lay in the future, and anyone who talked about sending men to the Moon was dismissed as an amiable eccentric. Yet there was a general interest in astronomy – after all, the sky is something which can hardly be overlooked – and my aim was to help the newcomer in making a start.

Today the overall situation is different. Astronomy and space research have become of everyday as well as academic interest, and it is not often that a year passes without its quota of spectacular discoveries. Great new telescopes have been built, the first space-stations have been put into orbit, and even photography is being rapidly superseded by electronic aids. Astronomy has become a fast-moving science instead of a comparatively static one.

Inevitably, this has affected the amateurs. In 1957 they were concerned mainly with the Solar System, and this was even more true when I started out at the early age of six (which takes me back to 1929). Modern amateurs often use highly sophisticated, computerized equipment and elaborate electronics. I must admit that in this respect I have not moved with the times. My main interest has always been in the Moon and planets, and my observatory is neither computerized nor electronically fitted. I still make my observations at the eye-end of my telescope – and after all, there is no better way to begin.

I have long since lost count of the number of letters I have had asking me, in various ways, how to start making a hobby out of astronomy. My answer is always the same. Do some reading, learn the basic facts, and then take a star-map and go outdoors on the first clear night so that you can begin learning the various stars and constellation patterns. The old cliché that 'an ounce of practice is worth a ton of theory' is true in astronomy, as it is in everything else.

At least I have one qualification: over the past years I suppose I have made almost every mistake that it is possible to make, and I hope, therefore, that I may be able to warn others against falling into the same traps. I have made no attempt to discuss activities which I do not pursue myself. If you want to find out about computer drives, photoelectric photometers, CCDs and the like, you must consult a book written by someone who knows more about them than I do.

It is fair to say that astronomy is still just about the only science in which the amateur can make valuable contributions today, and in which the work is welcomed by professionals. For example, amateurs search for new comets and 'new stars' or novæ, and since they generally know the sky much better than their professional colleagues they have a fine record of success. Routinely, they keep watch on objects such as variable stars, and they monitor the surfaces of the planets in a way that professionals have neither the time nor the inclination to do.

Quite apart from this, astronomy is a hobby which can be enjoyed by everyone. Take it up, and you will meet many people with the same interests; you will make many new friends, and, if you like, you can carry out some useful research. If you decide to follow it through, I can assure you that you will not regret it.

1

The Unfolding Universe

A subject can always be better understood if something is known about its history. Though we no longer worship our 'honourable ancestors', it is a distinct help to look back through time in order to see how knowledge has been built up through the centuries. This is particularly true with astronomy, which is the oldest science in the world — so old, indeed, that we do not know when it began.

Most people of today have at least some knowledge of the universe in which we live. The Earth is a globe nearly 8000 miles in diameter, and is one of nine planets revolving round the Sun. The best way of summing up the difference between a planet and a star is to say that the Earth is a typical planet, while the Sun is a typical star.

Five planets — Mercury, Venus, Mars, Jupiter and Saturn — were known to the ancients, while three more have been discovered in relatively modern times. Jupiter is the largest of them, and its vast globe could swallow up more than a thousand bodies the volume of the Earth, but even Jupiter is tiny compared with the Sun. The stars of the night sky are themselves suns, many of them far larger and more brilliant than our own, and appearing small and faint only because they are so far away. On the other hand, the Moon shines more brilliantly than any other object in the sky apart from the Sun. Appearances are deceptive; the Moon is a very junior member of the Solar System, and it has no light of its own. It has a diameter only about one-quarter that of the Earth, and it is much the closest natural body in the sky.

The whole celestial vault seems to revolve round the Earth once in 24 hours. This apparent motion is due, of course, to the fact that the Earth is spinning on its axis from west to east. Of all the celestial objects, only the Moon genuinely moves round the Earth.

We are used to taking these facts for granted, but in early times it was (rather naturally) believed that the Earth was flat and stationary. The Sun and Moon were worshipped as gods, and the appearance of anything unusual, such as a comet, was taken to be a sign of divine displeasure.

It is usually said that the first astronomers were the Chaldæans, the Egyptians and the Chinese. In a way this is true enough; these ancient civilizations made useful records, but they had no real understanding of the nature of the universe or even of the Earth itself.

The main story begins around 3000 BC, when the 365-day year was first adopted in

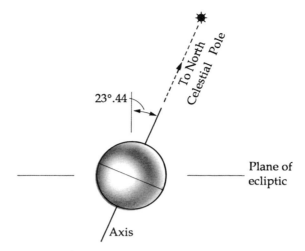

Fig. 1.1. Inclination of the Earth's axis.

Egypt and in China. This, too, was the approximate date of the building of that remarkable structure which we know as the Great Pyramid of Cheops, still one of the world's main tourist attractions. Astronomically, it is of special interest because its main passage is aligned with what was then the north pole of the sky.

The Earth's axis of rotation is inclined at an angle of 23½ degrees to the perpendicular to its orbit, and points northward to the celestial pole (Fig. 1.1). Today the pole is marked approximately by a bright star known as Polaris, familiar to every navigator because it seems to remain almost stationary while the entire sky revolves round it. In Cheops' time, however, the polar point was in a different position, close to a much fainter star, Thuban in the constellation of the Dragon. The reason for this change is that the Earth is 'wobbling' slightly, in the manner of a gyroscope which is running down, so that the direction of the axis is describing a small circle in the sky. The effect is very slight, but the shift of the pole has become appreciable since the Pyramid was built.

The Egyptians divided up the stars into constellations, though their scheme was different from that which we follow today. They also made good measurements, and they regulated their calendar by the 'heliacal rising' of the star Sirius – that is to say, the date when Sirius could first be seen in the dawn sky. Some of their other ideas were very wide of the mark. They believed the world to be rectangular, with Egypt in the middle, and that the sky was formed by the body of a goddess with the rather appropriate name of Nut.

The Chinese were equally good observers, and made careful records of comets and eclipses. Total eclipses of the Sun are particularly spectacular, and at this point I cannot resist re-telling a famous legend, even though experts assure me that it is certainly untrue! Here, then, is the story of Hsi and Ho.

The Moon revolves round the Earth once a month, while the Earth takes a year to complete one journey round the Sun. The Moon is much smaller than the Sun, but it is also much closer, so that — by pure chance — the two look almost exactly the same size. When the Sun, Moon and Earth move into an exact line, with the Moon in the mid-position, the result is a total solar eclipse. The Moon blots out the bright disk of the Sun, and for a few moments — never as long as eight minutes — we can see the glorious pearly corona and the 'red flames' or prominences; the sky becomes so dark that stars can be seen.

The Chinese knew how to predict eclipses — more or less — but they did not know that the Moon was involved; they thought that the Sun was in danger of being eaten by a hungry dragon, so that the only course was to scare the beast away by shouting, screaming, wailing, and beating gongs and drums. (It always worked!) The legend says that in 2136 BC, during the reign of the Emperor Chung K'ang, the Court Astronomers, Hsi and Ho, failed to give due warning that an eclipse was due, so that no preparations were made — and since Hsi and Ho had imperilled the whole world by their neglect of duty, they were summarily executed. I am sorry that the experts have demolished this tale. Had it been true, Hsi and Ho would have been the first known scientific martyrs in history.

It was the Greeks who turned astronomy into a true science, because they not only made observations but also tried to interpret them. The first of the great philosophers was Thales of Miletus, who was born in 624 BC; the last was Ptolemy of Alexandria, and with his death, in or about 180 AD, the classical period of science came to an end. The whole of the Greek period spread over eight centuries, so that, in time, Ptolemy was as far away from Thales as we are from the Crusades.

Thales himself probably believed the Earth to be flat, but unfortunately all his original writings have been lost, and for definite arguments against the flat-earth theory we must turn to Aristotle, who lived from 384 to 322 BC. Aristotle pointed out that the stars change in altitude above the horizon according to the latitude of the observer. For example, Polaris is fairly high in the sky of Greece, because Greece is well north of the Earth's equator; from Egypt, it is lower; from southern latitudes it can never be seen at all, because it never rises above the horizon. On the other hand Canopus, a brilliant star in the southern hemisphere of the sky, can be seen from Egypt but not from Greece. This is just what would be expected on the theory of a round Earth, but is quite impossible to explain if we assume the world to be flat. Aristotle also noticed that when the Earth's shadow falls across the Moon, the edge of the shadow is curved, indicating that the surface of the Earth must also be curved.

The next major step was taken by Eratosthenes of Cyrene, who succeeded in measuring the length of the Earth's circumference. His method was most ingenious, and proved to be remarkably accurate. Eratosthenes was in charge of a great scientific library at Alexandria, Egypt, and from one of the books available to him he learned that at the time of the summer solstice, the 'longest day' in northern latitudes, the Sun was vertically overhead as seen from the town of Syene (the modern Assouan), some distance up the Nile. At Alexandria, however, the Sun was then seven degrees from the zenith or overhead point (Fig. 1.2). A full circle contains 360 degrees, and seven is about $\frac{1}{50}$ of 360, so that if the Earth is spherical its circumference must be 50 times the distance from Alexandria to Syene. Eratosthenes may have arrived at a final value of

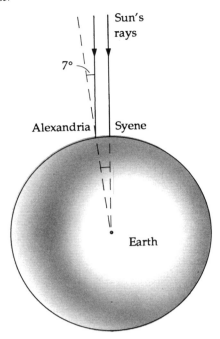

Fig. 1.2. Eratosthenes' method of measuring the circumference of the Earth. When the Sun was overhead at Syene, it was 7½ degrees from the vertical as seen from Alexandria.

24 850 miles, which is only 50 miles too small. Admittedly there is some doubt about this, because we are not sure of the length of the unit he was using, but in any case he was not very wide of the mark. His estimate was much better than the value adopted by Christopher Columbus during his voyage of discovery so many centuries later, which partly explains why Columbus came home without having any real idea of where he had been!

If the Greeks had taken one step more, and put the Sun in the centre of the planetary system rather than the Earth, the subsequent history of astronomy would have been very different. Some of the philosophers, such as Aristarchus of Samos, did believe the Earth to be in orbit round the Sun, but they could give no proof, and the later Greeks went back to the idea of a motionless, all-important Earth.

Much of our knowledge of ancient astronomy is due to one man, Ptolemy of Alexandria (more properly, Claudius Ptolemæus) who flourished between around 120 and 180 AD. We know absolutely nothing about his personality, but he was undoubtedly a brilliant observer as well as an expert theorist, and recent attempts to discredit him have been singularly unsuccessful. His main work has come down to us by way of its Arab translation, the *Almagest*. It is really a summary of ancient science, and it describes the theory of the central Earth which Ptolemy had perfected.

On the Ptolemaic pattern, all the celestial bodies move round the Earth. The Moon is closest; then come Mercury, Venus, the Sun, Mars, Jupiter, Saturn and finally the

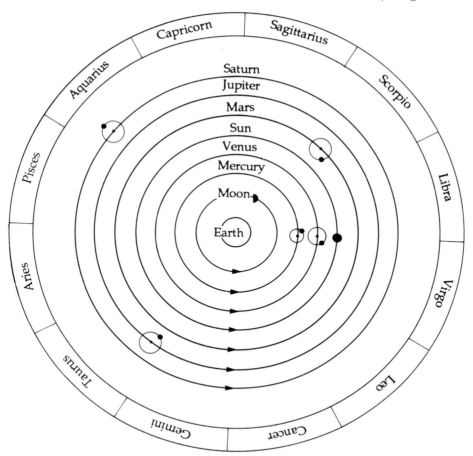

Fig. 1.3. The Ptolemaic system.

stars. Ptolemy maintained that since the circle is the 'perfect' form, and nothing short of perfection can be allowed in the heavens – all bodies must move in strictly circular paths. Unfortunately the planets have their own way of behaving. Ptolemy was an excellent mathematician, and he knew quite well that the planetary motions cannot be explained on the hypothesis of uniform circular motion round a central Earth. His answer was to work out a complex system according to which each planet moves in a small circle or 'epicycle', the centre of which – the 'deferent' – itself moves round the Earth in a perfect circle (Fig. 1.3). As more and more discrepancies came to light, more and more epicycles had to be introduced, until the whole system had become hopelessly artificial and cumbersome. Yet it did fit the observations – and since Ptolemy knew nothing about the nature of gravitation, he could hardly have done any better.

Hipparchus of Nicæa, who had lived two centuries before Ptolemy, had drawn up a detailed and accurate star catalogue. The original has been lost, but Ptolemy reproduced it in the *Almagest*, with additions of his own, so that most of the work has come down to us. We still use the 48 constellation figures which Ptolemy described, even though their boundaries have been modified and new groups added.

When the Greek period came to an end, all scientific progress came to an abrupt halt. The great library at Alexandria was looted and burned in 640 AD, by order of the Arab Caliph Omar, though most of the books may have been scattered earlier; in any case, the loss of the Library was irreparable, and scholars have never ceased to regret it. When interest in the skies did return, it came – ironically enough – by way of astrology.

Even today, there are still some people who do not know the difference between astronomy and astrology. In fact, the two are completely different. Astronomy is an exact science; astrology is a relic of the past, and there is no scientific or rational basis for it, though it still has a considerable following – not only in countries such as India, but in Europe as well.

The best way to define astrology is to say that it is the superstition of the stars. Each celestial object is supposed to have a definite influence upon the character and destiny of each human being, and by casting a horoscope, which is basically a chart of the positions of the planets at the time of the subject's birth, an astrologer claims to be able to look into the future. It takes one back to the Dark Ages, and when an astrologer is asked 'why' he believes that this sort of procedure works, he may well be honest enough to reply that he has no idea. Obviously some astrological predictions are correct – as was once said by an eminent judge, 'it is impossible always to be wrong' (though some modern politicians do their best). My own comment is that astrology does prove one fact, i.e. that 'there's one born every minute'. At least it is fairly harmless so long as it is confined to pier heads, circus tents, and the less serious columns of the Sunday papers.

However, mediæval astrology did at least lead to a revival of true astronomy, because the Arabs, who led the way, needed accurate star catalogues in order to cast their horoscopes. They also needed a knowledge of the movements of the Moon and planets. There were even observatories – very different from the domed buildings of today, but observatories none the less.

The main handicap was still the universal belief in the Ptolemaic theory. So long as men refused to believe that the Earth could be moving round the Sun, it was almost impossible to make any real advance. Things were not helped by the attitude of the Church, which in those days was all-powerful. Any criticism of Ptolemy was regarded as heresy. Since the usual fate of a heretic was to be burned at the stake, it was clearly unwise to be too candid.

The first serious signs of the approaching struggle came in 1546, with the publication of *De Revolutionibus Orbium Cælestium* (Concerning the Revolutions of the Celestial Orbs) by a Polish canon, Mikołaj Kopernik, better known to us as Copernicus. Copernicus was a clear thinker as well as being a skilful mathematician, and at a fairly early stage in his career he saw so many weak links in the Ptolemaic system that he felt bound to question it. It seemed unreasonable to believe that the stars could circle the Earth once a day. In his own words, 'Why should we hesitate to grant the Earth a motion natural and corresponding to its spherical form? And why are

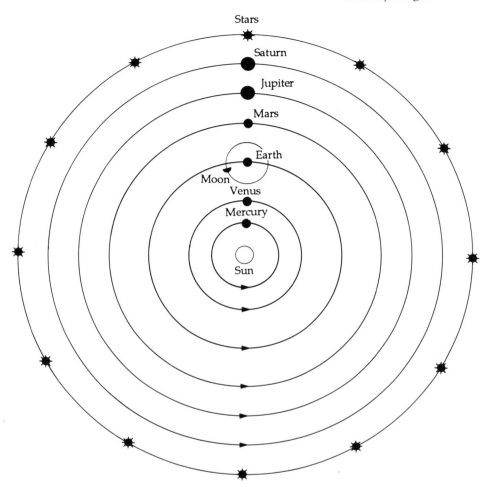

Fig. 1.4. The Copernican system.

we not willing to acknowledge that the *appearance* of a daily rotation belongs to the heavens, its *actuality* to the Earth?' The relation is similar to that of which Virgil's Æneas said, 'We sail out of the harbour, and the countries and cities recede.'

Copernicus' next step was even bolder. He saw that the movements of the Sun, Moon and planets could not be explained by the old system even when Ptolemy's various circles and epicycles had been allowed for, and so he rejected the whole theory. He placed the Sun in the centre of the system (Fig. 1.4), and relegated the Earth to the status of an ordinary planet.

It is fair to say that this was Copernicus' only major contribution; he made many errors, and in particular he retained the idea of perfectly circular orbits, so that in the end he was even reduced to bringing back epicycles. Still, he had taken the essential step. He was wise enough to be cautious; he knew that he was certain to be accused of heresy, and though his theory was more or less complete by around 1530 he refused to publish it until the year of his death. As he had foreseen, the Church was bitterly hostile; one of his strongest critics was Martin Luther. Later, in 1600, the Italian philosopher Giordano Bruno was actually burned in Rome because he insisted that Copernicus had been right. True, this was not Bruno's only crime in the eyes of the Inquisition, but it was certainly a serious one.

Tycho Brahe, born in Denmark only a few months after Copernicus died, was a man of entirely different type. He was a firm believer in astrology, and an equally firm opponent of the Copernican system, preferring a sort of hybrid pattern according to which the planets moved round the Sun while the Sun itself moved round the Earth. He built an observatory on the island of Hven, in the Baltic, and between 1576 and 1596 he made thousands of amazingly accurate observations of the positions of the stars and planets, finally producing a catalogue which was far better than those of Ptolemy or the Arabs. Of course he had no telescopes, but his measuring instruments were the best of their time, and Tycho himself was a magnificent observer.

The story of his life would need a complete book to itself. Tycho is, indeed, one of the most fascinating characters in the history of astronomy. He was proud, arrogant and grasping, with a wonderful sense of his own importance; he was also landlord of Hven, and the islanders had little cause to love him. His observatory was even equipped with a prison for those who refused to pay their rents; he had a false nose, to replace the original which had been sliced off in a student duel, and his retinue included a pet dwarf. Yet despite all his shortcomings, he must rank with the intellectual giants of his age, and Hven became very much a scientific centre; one man who visited it was the King of Scotland, afterwards James I of England. After Tycho left the island, following a quarrel with the Danish court, his observatory, Uraniborg, fell into ruin, and today nothing remains of it. I visited it a few years ago; the site is covered by grass, though it is overlooked by a huge statue of Tycho himself.

When Tycho died, in 1601, his observations fell into the hands of his last assistant, a young German named Johannes Kepler. After years of careful study, using Tycho's observations of the planets (particularly Mars), Kepler saw that the movements of the planets could be explained neither by circular motion round the Earth nor by circular motion round the Sun, so that there was something badly wrong with Copernicus' system as well as with Ptolemy's. Finally he found the answer. The planets do indeed move round the Sun, but not in perfect circles. Their paths, or orbits, are elliptical.

One way to draw an ellipse is shown in Figure 1.5. Fix two pins in a board, and join them with a thread, leaving a certain amount of slack. Now loop a pencil to the thread, and draw it round the pins, keeping the thread tight. The result will be an ellipse,* and the distance between the two pins or 'foci' will be a measure of the eccentricity of the ellipse. If the foci are close together, the eccentricity will be small, and the ellipse very little different from a circle; if the foci are widely separated, the ellipse will be long and narrow. Obviously, a circle is simply an ellipse with an eccentricity of zero.

The five planets known in Kepler's day proved to have orbits which were almost circular, *but not quite*. The slight departure from perfect circularity made all the

---

* The method is excellent in theory. In practice, what usually happens is that the pins fall down or the thread breaks. One day, I hope to carry out the whole manœuvre successfully!

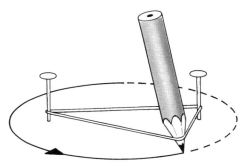

Fig. 1.5. How to draw an ellipse.

difference, and Tycho's observations fell beautifully into place, like the last pieces of a jigsaw puzzle. The age-old problem had been solved, though the Church authorities were not in the least impressed, and kept up their opposition for as long as they could. Kepler's three Laws of Planetary Motion, the last of which was published in 1618, paved the way for the later work of Isaac Newton.

Kepler's work was not the only important development of the early years of the seventeenth century. In 1608 Hans Lippershey, a spectacle-maker of Middleburg in Holland, found that by arranging two lenses in a particular way he could obtain magnified views of distant objects. Spectacles had been in use for some time, but nobody had hit upon the principle of the telescope until Lippershey did so, apparently by accident.

A refracting telescope consists basically of two lenses. One, the larger, is the object-glass; its function is to collect the rays of light coming from the target object, and bunch them together to form an image at the focus (Fig. 1.6). The image is then enlarged by a smaller lens or eyepiece.

The news of the discovery spread across Europe, and came to the ears of Galileo Galilei, Professor of Mathematics at the University of Padua (also, incidentally, the real founder of the science of experimental mechanics). Galileo was quick to see that the telescope could be put to astronomical use, and 'sparing neither trouble nor expense', as he put it, he built an instrument of his own. It was a tiny thing, much less effective than modern binoculars, but it helped toward a complete revolution in scientific thought.

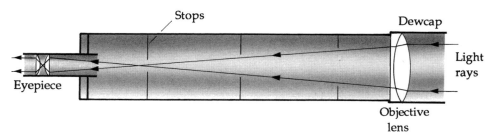

Fig. 1.6. Principle of the refractor.

Galileo first used his telescope in January 1610, and proceeded to make a whole series of spectacular discoveries. The Moon was found to be covered with dark plains, lofty mountains and giant craters; Venus, the Evening Star of the ancients, presented lunar-type phases, so that it was sometimes crescent, sometimes half and sometimes nearly full; Jupiter was attended by four smaller bodies or satellites, while the Milky Way proved to be made up of innumerable stars.

Galileo had always believed in the Copernican theory, and his telescopic work made him even more certain. For example, Venus could never show phases if it moved according to the Ptolemaic system, while the satellites of Jupiter proved that there must be more than one centre of motion in the Solar System. Unlike Copernicus, Galileo was headstrong and impetuous, and he found himself in serious trouble with the Church. He was accused of heresy, brought to trial in Rome, and forced to 'curse, abjure and detest' the false theory that the Earth moves round the Sun. But few people outside the Church were deceived, and before the end of the century the Ptolemaic theory had been abandoned forever. The publication of Newton's *Principia*, in 1687, marked the end of what is often called the Copernican Revolution.

Most people have heard the story of Newton and the apple. It is interesting because unlike most stories of similar type, such as Canute and the waves, it is probably true. Apparently Newton was sitting in his garden one day when he saw an apple fall from its branch to the ground, and realized that the force pulling on the apple was the same force as that which keeps the Moon in its path round the Earth. From this he was led on to the concept of universal gravitation, upon which the whole of later research has been based. It is fair to say that Kepler found out 'how' the planets move; Newton discovered 'why' they do so.

Newton also constructed an entirely new type of telescope. Galileo's instrument was a refractor, using an object-glass to collect its light. Newton believed that refractors could never be completely satisfactory, and he looked for some way out of the difficulty. Finally he decided to do away with object-glasses altogether, and to collect the light by means of a specially-shaped mirror.

When Newton rejected the refractor as unsatisfactory, he was making one of his rare mistakes. However, the Newtonian 'reflector' soon became popular, and has remained so. Mirrors are easier to make than lenses, and today all the world's largest telescopes are of the reflecting type.

Astronomy was growing up. So long as observations had to be made with the naked eye alone, little could be learned about the natures of the planets and stars; their movements could be studied, but that was all. As soon as telescopes made their appearance, true observatories were built. Copenhagen and Leyden took the lead; the Paris Observatory was completed in 1671, and Greenwich in 1675.

Greenwich was founded for a special reason. England has always been a seafaring nation; and before the development of reliable clocks, the only way in which sailors could fix their position when far out in the ocean, out of sight of land, was to observe the position of the Moon against the stars. This involved the use of a good star catalogue, and even the best one available, Tycho's, was not accurate enough. King Charles II, that much-maligned monarch, therefore ordered that the star places must be anew examined and corrected for the use of my seamen'. A site was selected in the Royal Park at Greenwich, and Sir Christopher Wren, who was Professor of Astronomy at Oxford long before he turned over to architecture, designed the first buildings, which still stand. The Rev. John Flamsteed was appointed Astronomer Royal, and in due course the new star catalogue was completed, though – ironically – the invention of the marine chronometer meant that it was never used in quite the manner originally intended.

Telescopes continued to be improved. Some of the early refractors were curious indeed; one of them, used by the Dutch observer Christiaan Huygens, was over 200 feet long, so that the object-glass had to fixed to a mast. But gradually the worst difficulties were overcome, and both refractors and reflectors gained in power and convenience. Mathematical astronomy made equally rapid strides. The great obstacle here had always been the Ptolemaic system, and once that had been swept away the path was clear. The distance between the Earth and the Sun was measured with reasonable accuracy, and in 1675 the Danish astronomer Ole Rømer even measured the velocity of light, which proved to be 186000 miles per second. Rømer did this, incidentally, by observing the movements of the four bright satellites of Jupiter.

But though knowledge of the bodies of the Solar System had been improved out of all recognition, little was known about the stars, which were still regarded as mere points of reference. The first serious attack on stellar problems was made by William Herschel, who is rightly termed 'the father of stellar astronomy'.

Herschel was born in Hanover in 1738, eleven years after the death of Newton. He came to England, and became organist at the Octagon Chapel in Bath, but his main interest was astronomy, and he built reflecting telescopes which were the best of their age. The largest of Herschel's telescopes, completed in 1789, had a mirror 48 inches in diameter and a focal length of 40 feet. The mirror still exists, and now hangs on the wall of Flamsteed House in Greenwich, though it has not been used since Herschel's time.

Herschel had his living to earn, and for some years he could not afford to spend all his time following astronomical research. Then, in 1781, he made a discovery which altered his whole life. One night he was examining some faint stars in the constellation of Gemini, the Twins, when he came across an object which was certainly not a star. At first it was thought to be a comet, but as soon as its path was worked out there could no longer be any doubt about its nature. It was not a comet, but a new planet – the world we now call Uranus.

The discovery was quite unexpected. There were five known planets, and these, together with the Sun and the Moon, made a grand total of seven. Seven was the magical number of the ancients, and it had been thought that the Solar System must be complete. Herschel became world-famous; he was appointed Court Astronomer to King George III, and henceforth he was able to give up his musical career altogether.

Herschel set himself a tremendous programme. He decided to explore the whole of the sky, so that he could form some idea of the way in which the stars were distributed. Until the end of his long life, in 1822, he worked patiently at his task, and his final conclusions have proved to be reasonably accurate (Fig. 1.7). He was, for example, the

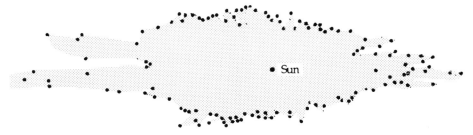

Fig. 1.7. Herschel's description of the shape of the Galaxy.

first to draw a fairly good picture of the shape of our star-system or Galaxy, which he described as being like 'a cloven grindstone' – though I prefer the less romantic idea of two fried eggs clapped together back-to-back!

Naturally, Herschel made thousands of discoveries during his sky-sweeps. Many apparently single stars proved to be double, and there were also clusters of stars, as well as faint luminous patches known as nebulæ, from the Latin word meaning 'clouds'. Herschel was a meticulous observer. He catalogued all his discoveries, and when we examine his published papers we can only marvel at the amount of work he managed to do. Yet some of his ideas were curious by modern standards; he believed that the Moon and planets must be inhabited, and went so far as to maintain that there were beings living in a cool, temperate region below the surface of the Sun.

Herschel lived in England for most of his life, and never saw the stars of the far south, which never rise here. It was fitting that the completion of his sky-sweeps should be accomplished later by his son, Sir John Herschel, who travelled to the Cape of Good Hope, taking a powerful telescope with him, and remained there for several years.

Another famous observer of this period was Johann Schröter, Chief Magistrate of the little German town of Lilienthal. Unlike Herschel, Schröter concentrated mainly upon observations of the Moon and planets, and he is the real founder of selenography, the physical study of the lunar surface. Unfortunately Schröter's telescopes, together with all his unpublished work, were destroyed by the invading French armies in 1814, and Schröter himself died two years later.

In the early years of the nineteenth century a German optician, Joseph Fraunhofer, began to experiment with glass prisms. Newton had already found that ordinary 'white' light is not white at all, but a blend of all the colours of the rainbow. Fraunhofer realized that this discovery could be turned to good account, and his work led to the development of a new instrument, the astronomical spectroscope.

Just as a telescope collects light, so a spectroscope analyzes it. By studying the spectra produced, it is possible to find out a great deal about the substances present in the material which is emitting the light. For instance, the spectrum of the Sun shows two dark lines which can be due only to the presence of the element sodium, so that we can prove that sodium exists in the Sun. To the professional astronomer of today, spectroscopes are invaluable; by using them together with telescopic equipment, we can track down the elements present in the stars, and even in star-systems far away in the depths of space.

In 1838 Friedrich Bessel, Director of the Observatory of Königsberg, returned to the problem of the distances of the stars — a problem which had defeated even Herschel. By studying the apparent movements of 61 Cygni, a dim star in the constellation of the Swan, Bessel was able to show that it lay at a distance of about 60 million million miles. About the same time a Scottish astronomer named Thomas Henderson, who had been working at the Cape of Good Hope, measured the distance of the bright southern star Alpha Centauri, and arrived at the reasonably accurate value of about 20 million million miles. The real value is about 24 million million miles, so that Henderson underestimated somewhat. Alpha Centauri is a triple star, and the faintest member of the trio (Proxima) remains the nearest known body outside our own Solar System.

24 million million miles! Our brains are not able to appreciate such distances, and it is clear that the mile is too short to be a convenient unit of length. One might as well try to give the distance between London and New York in centimetres. Fortunately there is a much better unit available, based upon the velocity of light.

Light travels at 186 000 miles per second. A ray from the Sun takes 8.6 minutes to reach us, but for Alpha Centauri the time of travel is 4.3 years; we see the star not as it is now, but as it used to be 4.3 years ago. We say, then, that Alpha Centauri is 4.3 -light-years away, while the distance of 61 Cygni is nearly 11 light-years.

Bessel's success gives us an added idea of the real unimportance of the Solar System. Rather than quote strings of figures, it may be better to give a scale model. If we begin by making the Sun a two-foot globe, and putting it in the middle of London, the Earth will become a pea at a distance of 215 feet; Uranus, the outermost of the planets known in Bessel's time, will be a plum $\frac{4}{5}$ of a mile away. What about the nearest star? It will not be in London, or even in England; it will be some 10 000 miles away, in the middle of Siberia. We have learned a great deal since the days when the Earth was thought to be the hub of the universe. And by now, we know that even Alpha Centauri and 61 Cygni are on our doorstep by cosmical standards.

Astronomical photography began in the first half of the nineteenth century; within a hundred years it had taken over from visual observation for most branches of research. Meanwhile, Herschel's great 48-inch reflector had been surpassed. In 1845 the third Earl of Rosse, in Ireland, built a 72-inch. It was cumbersome and awkward to use, but it was by far the most powerful telescope then in existence, and Lord Rosse used it well. He studied the clusters and nebulæ which had been listed by Herschel and others, and found that although some of the nebulæ looked like patches of gas, others were made up of stars; even more interesting was the fact that some of the 'starry nebulæ' showed a spiral structure, like Catherine-wheels. For decades, the Rosse telescope was the only instrument in the world capable of showing the spiral forms of the nebulæ.

Alone, the telescope could never decided upon the nature of the nebulæ; the spectroscope was able to do so. In 1864 Sir William Huggins examined a faint nebula in the constellation of the Dragon, and found that it was composed not of stars, but of luminous gas.

We now know that the nebular objects are of two types. There are the gas-and-dust clouds, such as the Great Nebula in the Sword of Orion, which are stellar birthplaces,

Fig. 1.8. The Andromeda Galaxy – like our Galaxy , a spiral system. (Photograph by Bernard Abrams, 10-inch reflector.)

and which shine because of the stars mixed in with them. Beyond our Galaxy there is a vast gulf, and then we come to the external star-systems, lying at immense distances. One of the most famous of them is the Great Spiral in Andromeda (Fig. 1.8), which can be seen with the naked eye as a dim misty patch, and which has proved to be a galaxy in its own right, even larger than our own; its distance is 2.2 million light-years, so that we are seeing it as it used to be well before Man appeared on the Earth. Herschel and Rosse had suspected something of the sort, though the question was not finally settled until 1923.

Lord Rosse's 72-inch reflector did not retain its lead for long, but in the latter half of the nineteenth century the astronomical scene was dominated by the great refractors, of which the largest was the 40-inch at the Yerkes Observatory in America – still the most powerful refractor ever constructed. But there is a limit to the size of a usable refractor, and in our own time the lead has been taken by reflectors. The American astronomer George Ellery Hale – who had an amazing knack of persuading friendly millionaires to finance his schemes – master-minded first a 60-inch, and then a superb telescope with a mirror 100 inches across, set up on the summit of Mount Wilson in California. He then planned a 200-inch reflector. Sadly, he did not live to see it completed; it came into action in 1948, on the summit of Palomar Mountain – also in California – and even today it is exceeded in sheer size only by the 236-inch reflector built in the Soviet Union. We must admit, though, that the 236-inch has never really come up to expectations.

Building a giant mirror is extremely difficult, because everything has to be so amazingly accurate. There is also the problem of building a mounting which is

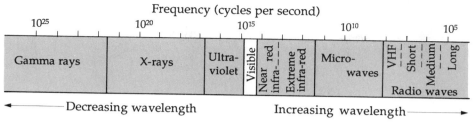

Fig. 1.9. *The electromagnetic spectrum.*

sufficiently stable and manœuvrable. Modern technology has come to the rescue, and the large telescopes of today are fully computer-controlled in a way which would have been quite impossible only a few years ago. The current emphasis is upon southern-hemisphere observatories; the main aim is to select the best possible sites for seeing conditions.

Also, there are telescopes of new design. The MMT or Multiple-Mirror Telescope at Mount Hopkins, Arizona, has not one mirror, but six – each 70 inches in diameter, working together and bringing the light from the target object to a common focus. Another development is to make a large mirror in sections, fitted together to make up the correct optical shape. Most major observatories are sited on the tops of high mountains, because the Earth's dirty, unsteady atmosphere is the astronomer's enemy; thus there are large telescopes on the summit of the extinct volcano of Mauna Kea, Hawaii, at an altitude of almost 14 000 feet, and in the Chilean Andes.

There are extra problems to be faced. The Mount Wilson Observatory, with its 60-inch and 100-inch reflectors, was temporarily closed in the 1980s because of light pollution from the neighbouring city of Los Angeles; the largest telescope ever built in Britain, the Isaac Newton Reflector, has been shifted from its site in Sussex to the clearer skies of La Palma, in the Canary Islands, while the original observatory in Greenwich Park has become a museum. Light pollution is something which cannot be ignored – as the average amateur astronomer knows only too well!

Until our own century, astronomers had to depend entirely upon the light-rays coming from space, but things are very different today. Light may be regarded as a wave-motion, and the colour of light depends upon its wavelength – from red (long) to violet (short) – but visible light makes up only a very small part of the total range of wavelengths, or 'electromagnetic spectrum' (Fig. 1.9). Beyond red we come to infra-red, microwaves and then to radio waves; beyond violet we find ultra-violet, X-rays, and finally the ultra-short gamma-rays.

In 1931 Karl Jansky, an American radio engineer of Czech descent, was carrying out investigations into 'static', on behalf of the Bell Telephone Company, when he found to his surprise that he was picking up long-wavelength radiations from the Milky Way. This was the start of the science of radio astronomy, which has now become of vital importance, and which is certainly not outside the scope of the skilful amateur.

Radio telescopes are not in the least like optical telescopes, and they do not produce visible pictures; one cannot look through them, as some people fondly imagine. The

Fig. 1.10. The 250-foot Lovell Telescope at Jodrell Bank, as I photographed it in 1988.

most famous example is certainly the Lovell Telescope at Jodrell Bank, Cheshire, England (Fig. 1.10), which has a collecting dish 250 feet in diameter. Then, too, we have telescopes designed mainly for collecting infra-red radiations, such as the UKIRT (United Kingdom Infra-Red Telescope) on Mauna Kea, where it has been joined by the James Clerk Maxwell Telescope, which concentrates upon the microwave region of the electromagnetic spectrum. But the Earth's air is always a handicap, because it blocks out many of the radiations coming from space, and the only solution is to 'go aloft'.

The Space Age began on 4 October 1957, when the Russians launched their first artificial satellite, Sputnik 1, which sped round the Earth sending out the famous 'bleep! bleep!' signals which will never be forgotten by anyone who heard them (as I did). Less than four years later Yuri Gagarin, of the Soviet Air Force, made the first manned trip beyond the Earth. Since then we have had artificial satellites of all kinds, both manned and unmanned, and we have sent men to the Moon. Robot probes to the planets have contacted all the members of the planetary family out as far as Neptune, and controlled landings have been made on Mars and Venus; there has been a concerted attack on Halley's Comet when, in 1986, no less than five space-craft by-passed it and one (the European Giotto) went right through its head; and it is no longer futuristic to talk about colonies on the Moon and expeditions to Mars. During the past 25 years we have learned more about the Solar System than we had been able to do throughout the whole of human history. The next great step should be the launching of a space telescope, named in honour of the great American astronomer Edwin Hubble. If all goes well, the Hubble Telescope should be in orbit very shortly, and operating from above the Earth's atmosphere it will be far more effective than any telescope can be if operating from ground level.

All this has caused a revolution in outlook, and it has been claimed that the day of the amateur observer is over. Yet nothing could be further from the truth. The amateur is more restricted than he used to be, and has had to become more specialized, but his rôle is as important as ever — perhaps more so. So do not be disheartened; even with the probes, the space-stations and the great mountain observatories, there is still invaluable work for the amateur to do.

2

Telescopes

When I was aged eleven – and that, I must admit, takes me back to the year 1934 – I acquired my first telescope. It was a 3-inch refractor, which I still have and which I still use. It cost the princely sum of £7 10s, which was a good deal in those days, and it meant that I had to save up pocket money, Christmas presents, birthday presents and everything else for what seemed a long time; but it was well worth it. Today, the same telescope would cost a great deal more then I paid for it; and when faced with a financial outlay of around £300 at least, many people sigh sadly and turn elsewhere.

I quite agree that telescopes are not cheap, and good, second-hand telescopes are now about as common as great auks. On the other hand, remember that the cost is non-recurring, and that a telescope will last a lifetime if it is even reasonably well-treated, whereas the photographer has to allow for constant expenditure on films, printing and other essentials. Also, it is quite interesting to compare the cost of a telescope with that of using public transport. A couple of rail journeys between (say) London and Edinburgh, or an air fare from New York to Los Angeles, will swallow up more money than you will pay for a satisfactory telescope.

On the other hand, there are definite traps for the unwary. What follows represents my own views, and not everyone will agree, but I am quite unrepentant.

As we have seen, telescopes are of two main types – refractors and reflectors. In a refractor, the light from the target object is collected by the object-glass, brought to focus, and the resulting image enlarged by the eyepiece. All the actual magnification is done with the eyepiece, and various eyepieces can be used with the same telescope. In a reflector of the Newtonian variety (so called because the first telescope of this kind was made by Sir Isaac Newton, and presented to the Royal Society in 1671), the light passes down an open tube and is collected by a curved mirror, after which it is sent up the tube on to a smaller, flat mirror; the flat mirror is placed at an angle, and sends the rays to the side of the tube, where they are brought to focus and the image is magnified by the eyepiece (Fig. 2.1). With a Newtonian, therefore, the observer looks into the side of the tube instead of up it.

The light-gathering power of the telescope depends upon the diameter of the object-glass (refractor) or main mirror (reflector). In my view, it is most unwise to buy any refractor with an object-glass less than 3 inches in diameter, or a Newtonian with a main mirror below 6 inches across. This is simply because the amount of light

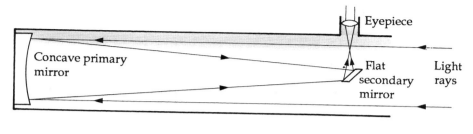

Newtonian

Fig. 2.1. Principle of the Newtonian reflector.

collected will be insufficient to allow for anything but a low magnification, and the field of view will probably be inconveniently small.

Unfortunately there are many small telescopes on sale — usually, though not always, Japanese — which look remarkably efficient, but are to all intents and purposes useless. In general, no telescope will bear a power of more than × 50 per inch of aperture, at best, so that a 3-inch will tolerate 150 (3 × 50) and so on. Dealers tend to exaggerate things beyond all reason. Not long ago I found an advertisement for a refractor which, it was claimed, could be used with a magnification of × 400. The aperture of the object-glass was $2\frac{1}{2}$ inches. This means that the absolute maximum you could expect would be × 125, and I doubt whether even this would be satisfactory, as it would involve really high-quality optics.

I have been sent many of these tiny telescopes to test, both refractors and reflectors (the 4-inch Japanese reflector is another trap). I have yet to find one which is of any real use, even if equipped with a rigid stand instead of the usual willowy mounting which quivers charmingly in the slightest breeze, so that the object you are trying to observe will describe a wild waltz across the field of view. Yet some of these telescopes cost many tens of pounds, and the buyer is certain to be disappointed, with the possible result that he will abandon astronomy altogether.

One typical advertisement which I saw recently ran as follows: 'Powerful astronomical telescope, × 400 and × 500, for only £180. Look at the craters of the Moon, the spots on the Sun and the stars of the Milky Way'. Nothing was said about the aperture of the telescope. It turned out to be a $2\frac{1}{4}$-inch refractor, giving a maximum usable power of little more than × 100. Neither was anything said about the risk of looking directly at the Sun — something else about which I hold strong views, and to which I will return in Chapter 5.

In short: avoid any telescope in which the dealer gives the magnification only. Remember, too, that an astronomical refractor (or Newtonian reflector) gives an inverted image. Obviously it is possible to add a correcting lens to turn the image the right way up again — but each time a ray of light passes through glass it is slightly weakened, and the astronomical observer is anxious to collect as much light as he can, so that the erecting lens is left out. After all, it does not in the least matter, and in most astronomical drawings and photographs the image is inverted.

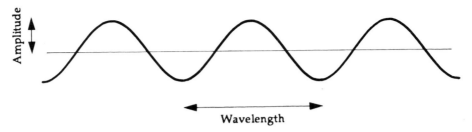

Fig. 2.2. *Wavelength (the distance between one wave-crest and the next).*

With a proper telescope, the object-glass will not be a single lens. As Newton found, more than three centuries ago, what we call 'white' light is really made up of all the colours of the rainbow, from red to violet. Light may be regarded as a wave motion, and the distance from one crest to the next is called the wavelength (Fig. 2.2). Red light has a longer wavelength than blue or violet, so that the object-glass does not bend it so much. The difference in the amount of bending or 'refraction' means that the red rays are brought to focus at a greater distance from the object-glass (Fig. 2.3). This causes trouble, and the image of a bright object, such as a star, will appear to be surrounded by false colour. These coloured rings may look very pretty, but to the astronomer they are most unwelcome. This is why good object-glasses are made up of several lenses, composed of different kinds of glass whose chromatic properties tend to cancel each other out. The effect can never be eliminated, but it can be very much reduced. Needless to say, the object-glass of a small, cheap refractor will usually be single, and the false-colour nuisance will be pronounced.

The distances between the object-glass and its focal point is known as the focal length, and this length, divided by the diameter of the object-glass, gives the focal ratio (usually abbreviated to f/ratio). For example, my 3-inch refractor has a focal length of 36 inches, so that the focal ratio is 36/3, or 12. The eyepiece combination (which is usually, though not always, compound) has its own focal length, and the magnification obtained depends on the ratio of the focal length of the eyepiece to that of the object-glass. With my f/12 refractor, an eyepiece of focal length $\frac{1}{2}$ inch will give a magnification of $36/\frac{1}{2}$, or × 72. With an object-glass of focal length 48 inches, the same eyepiece would give a magnification of $48/\frac{1}{2}$, or × 96.

You might therefore think that the way to get the best out of an eyepiece must be to use it with an object-glass of very long focal length. Unfortunately, things are not quite so simple as that. A large object-glass will collect more light than a smaller one. Suppose that I use a very short-focus eyepiece, say $\frac{1}{20}$ inch, on my 3-inch refractor? The magnification will be $36/\frac{1}{20}$, or × 720, but the image will be so faint that it will be practically invisible. The object-glass is too small to collect enough light to satisfy so powerful an eyepiece. If I want to use a magnification of × 720, I must use a larger telescope.

Making a lens is a task for the expert, and very few amateurs will feel like attempting it. If you need a refractor, the only sensible course is to buy it complete.

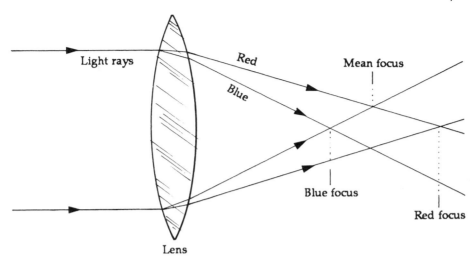

Fig. 2.3. *Unequal refraction of light. Short-wavelength light is more sharply refracted than long-wavelength light.*

Expense is the problem; a 3-inch is within the range of many people, but larger refractors are prohibitively costly. Yet there is an alternative, which I strongly recommend as a useful first step in any case. Rather than waste money on a very small telecope, buy a pair of binoculars, which have many advantages and relatively few drawbacks.

A pair of binoculars is nothing more nor less than two small refractors joined together, so that the observer can use both eyes instead of only one. The larger the aperture, the greater will be the amount of light which can be collected, and the brighter will be the image. Binoculars are classified according to their aperture and magnification. Thus 7 × 50 indicates a magnification of × 7, with each object-glass 50 millimetres across; 12 × 50 gives a magnification of × 12, and so on.

If you decide to buy a pair of binoculars, there are various things to consider. The higher the magnification, the smaller the field of view, and with any power above about × 12 the binoculars are bound to heavy, so that they will be difficult to hold steady and will have to be used with some sort of mounting. I recommend a power of between × 7 and × 12, with aperture 30–70 millimetres. For general viewing I find 7 × 50 very convenient. I also have 11 × 80 binoculars, which are of splendid quality and give a wide field, but are admittedly rather heavy to hand-hold.

Binoculars will give excellent views of the mountains and craters of the Moon, as well as endless star-fields, clusters, coloured stars and much else. They can be used for some real research, particularly with regard to variable stars and comet-hunting, and of course they are equally good for everyday use – bird-watching, looking at ships out to sea, and so on – whereas a small astronomical telescope is not. If you do want more powerful binoculars, such as 20 × 70, you will have to make or buy a stand; a converted camera tripod will serve, and nowadays it is also possible to buy chest-mountings which are quite helpful. (Remember one thing: as soon as you take the binoculars out of their case, put the safety-cord round your neck. Otherwise, sooner or later the binoculars will be dropped, with disastrous results.)

I will not go into further details here, because I have done so elsewhere,* but to sum up: if you follow my advice, you will start with binoculars, and avoid small telescopes. If you retain your interest, then it will be time to consider buying a telescope which will be of real and lasting value.

There is another advantage, too. Binoculars can be tested before being bought; all you have to do is to look through them. If the focus is soft, or there is any false colour, then the images are of inferior quality. I do not recommend buying any binoculars without actually trying them.

Now let us consider reflecting telescopes, which are of various types. With the Newtonian, the flat mirror blocks out some of the light from the main mirror or speculum, but the loss is not serious, and there is no way of avoiding it. On the credit side, a mirror reflects all wavelengths equally, so that there is no false-colour problem; colour estimates made with a reflector are always more reliable than those made with a refractor.

Mirrors are nearly always made of glass, coated with a reflecting layer of silver or aluminium. The layer has to be extremely thin. Silvering can be done at home, with care, but aluminizing involves special equipment. Fortunately, it is not expensive to have a mirror re-aluminized.

A reflector is classified according to the aperture of its main mirror, but we must be careful when comparing mirrors with lenses; inch for inch, the lens will give the better result. A 6-inch refractor is much more effective than a 6-inch reflector. Generally speaking, I would say that for most purposes a 4-inch refractor can match a 6-inch Newtonian, though opinions differ.

Most Newtonians have focal ratios of between f/7 and f/9. There are other optical systems, becoming more and more popular with amateurs, notably the Cassegrain, in which the secondary mirror is convex instead of flat, and is placed in front of the focus of the main mirror, so that the light is reflected back to the eyepiece through a hole in the centre of the speculum (Fig. 2.4). More complex designs, such as Maksutovs and Schmidts, combine lenses and mirrors. I do not propose to go into more detail here, because it is usual for the beginner to equip himself either with a refractor or with a Newtonian reflector. The more elaborate forms will come later.

The heart of a Newtonian is its main mirror, which has to be extremely accurate. If the curve is faulty, the images will be poor, and the telescope will be virtually useless. The main mirror is also the most expensive part of the telescope, but it is within the scope of a skilled amateur to make one. The principle of grinding a mirror into the correct optical curve is to use two glass disks, one of which will turn into the final mirror while the other is merely a 'tool'. The tool is fastened to a bench, and the mirror placed on top of it, with water and carborundum powder between the two. The mirror is then slid to and fro, while the operator rotates it and walks round the bench. The

---

* *Exploring the Night Sky with Binoculars* (Cambridge University Press, 1986).

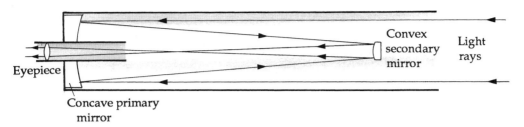

Fig. 2.4. Principle of the Cassegrain reflector.

tool will be worn away round its edge, and will become convex, while the mirror will be worn away toward the middle and will become concave. The process is straightforward enough, but it is very time-consuming, so be warned! The mirror has then to be figured and polished, and here a moment's carelessness will ruin many hours of work. Numerous tests have to be made, and I do not want to minimize the problems; there will be moments when you will feel like hurling the mirror on to the ground and stamping on it. But it can be done, and many amateurs have made their own reflectors at very modest cost.

If you intend to try, I suggest starting with a relatively small aperture – say 6 inches, or even 4 inches; one's second mirror is always much better than one's first. The largest amateur-made mirror that I know has a diameter of 24 inches, but most people will be satisified with something less ambitious. If you can make a good 12-inch, you are doing very well indeed.

Reflectors are much more temperamental than refractors. The optics have to be periodically re-silvered or (preferably) re-aluminized, and they are always liable to go out of adjustment. If you find that you are seeing blurred or distorted images, the likely reason is that the optics are not properly squared-on, but in general this is something which is easy to put right.

Making a mounting is a purely mechanical task. Fig. 2.5 shows the 12½-inch reflector which I have used for the past forty years. It is set upon what is termed an altazimuth stand, so that it can move freely either in *alti*tude or in *azimuth* (east–west). The tube rests in a cradle and is kept in position solely by its own weight. The cradle can be rotated, and the telescope can be swung up and down by sliding the rod. The top of the rod is fitted with a worm, so that by moving the wheel the telescope can be shifted very slightly up or down, while the handle, attached to a joint, gives a similar slight rotation of the whole telescope. These slight rotation mechanisms are termed 'slow motions', and for any telescope of over 6 inches aperture they are more or less essential.

Fig. 2.6 shows a much simpler mounting, this time for my 3-inch refractor. It is simply a tripod, so that the telescope can be moved in any direction. Slow motions are not always fitted, but they certainly make for easier observing.

The next drawing, Fig. 2.7, is included as an Awful Warning. It is that appalling

Fig. 2.5. My 12½-inch altazimuth reflector at Selsey.

Fig. 2.6. A 3-inch refractor, mounted on a simple tripod.

Fig. 2.7. A pillar-and-claw mounting for a telescope. Not to be recommended!

contrivance known as the Pillar and Claw, beloved of dealers but despised by observers. It looks nice, and is cheap, but it is about as steady as a blancmange. Anyone who buys a refractor on a pillar and claw will be wise to buy a rigid tripod, and consign the original mount to the dustbin.

The trouble about altazimuth mountings is that they have to be moved all the time, both up-or-down and east-to-west, so as to follow the target object as it shifts across the sky. It is amazing how obtrusive this shift is. Remember, if you are using of (say) × 100, you are also speeding-up the movement 100 times, and the object will pass out of the field of view very quickly. Slow motions are helpful, but it is irritating to have to fiddle continuously with both wheel and handle. To do so, and also make notes of your observations, is distinctly frustrating. I always feel that an altazimuth is ideal only for an observer who has three hands. This applies also to the increasingly-popular Dobsonian mounting (Fig. 2.8), which consists essentially of a telescope in a cradle which is rotated by a turntable.

The only real answer is to use what is termed an equatorial mounting. Here, the polar axis is pointed to the celestial pole and the telescope is mounted on an axis at right angles to it, so that only the east-to-west motion has to be made; the up-or-down movement looks after itself. There are various designs. One is the German (Fig. 2.9), in which the telescope is balanced by a counterweight; this is a favourite type, though personally I prefer the Fork (Fig. 2.10), in which the telescope swings between two arms mounted on the polar axis. There is also the English mounting, in which the polar axis is supported at both ends; it is extremely firm, but it takes up a great deal of room, and has the disadvantage that the telescope cannot be pointed toward the region near the celestial pole. All these mountings can be home-made, and there is no real problem in adding a driving mechanism which will turn the telescope so as to keep pace with the revolving sky.

Always bear in mind that if you set out to make an equatorial mount, it must be really rigid. A good rule is to work out the maximum possible weight which it ought to be — and then multiply by three.

If you want to use the telescope for astronomical photography, an equatorial mounting is more or less essential, because you will have to give time-exposures; and if the telescope is not driven, even a few second's exposure will result in an unacceptable amount of blurring. Of course, you can also mount an ordinary camera on the telescope (Fig. 2.11), and use the telescope itself as a guide.

Many reflectors have skeleton tubes. These have the advantage of portability and lightness, and there is no trouble from the internal air-currents which plague a closed tube. On the other hand, skeletons tend to be delicate, and unless housed in an observatory they are affected by any stray light pollution. My 15-inch reflector is partly enclosed, and I feel that this has major advantages; there is no air-current trouble, and 'extras' such as finders can be attached to the solid part of the tube — which in this case is wooden.

Each type of telescope has its merits and drawbacks. Refractors are easy to handle, and need virtually no maintenance; also they give beautifully sharp, crisp images (always provided that the optics are good). On the other hand they are more

Fig. 2.8. A Dobsonian mounting – Nigel Bannister's 10-inch reflector.

Fig. 2.9. My 5-inch refractor, on a German-type mounting.

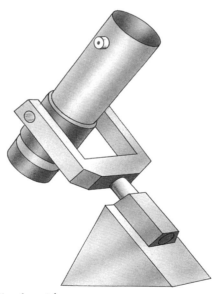

Fig. 2.10. A fork mounting for a telescope.

Fig. 2.11. Recommended method of mounting a camera on a telescope.

expensive than reflectors of equivalent light-grasp, and if mounted in the conventional way they can be awkward; if the target object is high in the sky, the eyepiece will be low down, and unless an angled prism is used at the eye-end the observer has to be something of a contortionist.

It is always wise to be on the alert when buying a telescope, even from a reputable dealer; it has to be admitted that some professionally-made telescopes are of poor quality. Be particularly careful with reflectors. The telescope may look inviting, with polished fittings and beautifully-painted tube; but if the mirror is poor, the performance also will be poor, and defects in a mirror do not always show up at first sight. Of course, one way of deciding is to make a practical test on a star image; but if the telescope is unmounted, or if it needs adjusting, this test may not be possible, in which case the only real safeguard is to seek advice from someone who has a sound knowledge of optics. (Local astronomical societies can often help.) The beginner who spends a large sum of money upon a telescope, only to find that the optics are faulty, is unlikely to receive much sympathy — and neither will he deserve it.

Let us assume, then, that you have acquired a telescope. What care must be taken of it, and what extra equipment is needed?

One addition is simplicity itself. A small sighting telescope or finder can be fitted (Fig. 2.12), and will be invaluable. Even a toy telescope will do, if nothing else is available. The finder must have a wide field of view, and it will save much time when you are trying to point the main telescope in the right direction — particularly with a Newtonian reflector, because it is not easy to squint 'up' the tube. The target is

Fig. 2.12. Typical examples of finders for telescopes.

brought to the centre of the finder field; if the adjustments are correct, it will then also be in the field of the main telescope. My own 15-inch has four finders, on the principle that the finder you want to use is always the one you can't get at. (I also have a revolving section on the upper end of the tube, so that the eyepiece can usually be brought to a convenient position; but with a smaller telescope, such as a 6-inch, this is not so important.)

With a refractor, a dew-cap is useful; this is simply a short tube which fits over the object-glass end of the telescope to prevent dust, dirt and dew from settling on the lens. It can be made from a cocoa-tin, or something of the sort, lined with blotting-paper to absorb the moisture. A solid cap should always be kept over the object-glass when the telescope is not actually in use.

If the object-glass needs cleaning, it should be brushed very gently with a camel's-hair brush and then wiped even more gently with a piece of silk or wash-leather. To take the various components of an object-glass apart is most unwise unless the owner has a really good idea of what he is doing. When some major adjustment does become necessary, it is well worth while to take the whole instrument to an expert.

Reflectors need more attention. As I have said , the main mirror and the flat need periodical re-coating with silver or aluminium; if you live on the coast (as I do), there is something to be said for having the mirror coated with rhodium. This is less reflective than silver or aluminium, but it is far more durable, and the mirror can even be washed. The main drawback is that rhodium cannot be removed without having the mirror re-figured, so make quite sure that you are happy about it before taking the plunge.

Both mirror and flat should be kept covered with a protective cap except when in use. Another word of warning may be timely here. Before using the telescope, uncap the flat before exposing the main mirror. I know of one luckless observer who uncovered the main mirror first — and then dropped the flat cover on to it. He spent the next few months in re-figuring his mirror.

Eyepieces are vitally important, since using a good telescope with a poor eyepiece is like using a good record-player with a bad needle. Theoretically, eyepieces are made to a standard thread, so that any eyepiece should fit any telescope; but remember that the magnification depends upon the focal length of the object-glass or mirror, so that an eyepiece which gives, say, × 60 on a 3-inch will not necessarily give × 60 on a 6-inch reflector. Japanese eyepieces are common nowadays, and many of them are excellent, but generally they have a different thread, so that a converter is needed. Fortunately this presents no real problem, because a converter is cheap and easy to buy or make.

It is advisable to have at least three eyepieces. One should give a low magnification, for star-fields and for general use; the second, higher magnification for more detailed views of planets and some stellar objects; the third, high magnification for use on good nights. For my 3-inch, f/12 refractor I have found that suitable magnifications are 36, 72 and 144; for a 6-inch reflector the corresponding powers might be 50, 120 to 180, and 250 to 300. Individual observers are bound to have their own ideas about this.

Do not try to use too high a power. If the image becomes even slightly blurred, change at once to a lower magnification. It may sound impressive to record that an observation has been made with ' × 400' or ' × 500', but it will usually be found that a smaller, sharper picture will be much more satisfactory. There are, incidentally, attachments known as Barlow lenses which can be used in conjunction with ordinary eyepieces to increase the magnification. Personally I dislike Barlows, but many people are enthusiastic about them.

When you have passed beyond the beginner's stage, you may well want to extend your range by using electronic or spectroscopic equipment, and of course there is immense pleasure to be gained from astronomical photography. I do not propose to go into detail here, because I am doing no more than describing the basic needs, but there are many books which will act as guides.

Summing up: If a telescope is to be bought, consider everything carefully before spending much money. Avoid very small telescopes (below 3-inches aperture for refractors, below 6 inches for Newtonian reflectors); they will not satisfy you for long, if at all. Never buy a telescope until you have taken an expert opinion on it, since although it may look quite sound it may prove to have faulty optics. Most important of all, do not trust your own judgement unless you are sure that you are really competent. It is only too easy to make a mistake.

3

Observatories – and Observing Hints

The amateur observer has a good many problems to face. People who live in towns, or close to towns, have to fight what is often a losing battle against light pollution and industrial haze; the glare of street lamps and shopping centres will cast a glow over the entire sky, effectively blotting out all but the brightest stars. Unfortunately there is no answer to this except to take one's telescope into the country, where the skies are dark. Portable telescopes must necessarily be fairly small; I suppose that the limit for convenient portability is 4 inches aperture for a refractor and 6 inches for a Newtonian reflector, though more complex optical systems are also more compact, and nowadays it is possible to buy 8-inch or even 12-inch reflectors of the Maksutov type which can be carried around easily. Their main drawback is that they are very expensive.

I admit that I have a strong dislike of moving telescopes around any more often than is absolutely necessary. It is only a matter of time before something gets dropped; and if the mounting is equatorial, there is always the problem of aligning the polar axis correctly with the celestial pole. Neither is it a good idea to leave a telescope permanently in the open, because mountings begin to rust or (if wooden) decay, and optics are not improved by constant exposure to damp. Obviously it is better to set-up the telescope inside some sort of observatory.

Most people are familiar with the law which is sometimes termed Murphy's Law and sometimes (in astronomy) as Spode's Law. It states, broadly, that if things *can* be awkward, they *are*. Even if you have a large garden, it always happens that the house is in the wrong place, cutting out that region of the sky which interests you most. Any trees will be in the most inconvenient positions possible.* The only solution is to site your observatory as advantageously as you can.

Make sure that you do your best to protect yourself from stray light. (I am lucky; I live in the end of a peninsula, and I have no marked sky-glow apart from a certain amount of what I call 'aurora Bognor Regis' in the north.) Street lights can sometimes

---

* If the trees are your own, of course, something can be done. At my home in Selsey there used to be a tall, stately pear-tree close to the site where I set up my main observatory; the tree looked nice, but it never produced any pears. One night it blocked my view of Saturn. Next day it turned into a small, stumpy pear-tree – and it now produces plenty of pears, so that clearly it got the message!

Fig. 3.1. Run-off shed for my 12½-inch reflector. The shed is in two parts, running back on rails.

be shielded – the run-off shed of my 12½-inch reflector has a screen to block out the light from the only lamp down my country lane – but this is not always easy, and again there is not much that can be done except make the best of it.

The essential need is for a very firm telescope mounting. Unless it is really rigid, it will be a source of permanent frustration. If you have a refractor, mount it upon a pillar which is fixed in concrete, so that it will not shift by even a fraction of an inch. The same applies to a reflector, though a pillar is not always needed. My 12½-inch Newtonian altazimuth has a wooden tripod which is simply bolted to a concrete platform; I have used it for over 40 years, and it has given no trouble at all.

Incidentally, the much-vaunted 'roof-top observatory' should be avoided at all costs. If you happen to have a house with a flat roof, resist the temptation to mount your telescope upon it. If you are above a heated dwelling, warm air swirling up will ruin the seeing. Neither is it practicable to pole a telescope through a bedroom window, for the same reason – apart from other hazards; I well remember that when I adopted this procedure, when I was very young, I dropped an eyepiece 30 feet on to the path below. Site your observatory as far away from your house as you can.

The classic form of observatory is that of a graceful dome, the upper part of which can revolve. I agree that this looks elegant, but it is not easy to make and is costly to buy. An alternative is the run-off shed (Fig. 3.1). Here, the 'observatory' is mounted on rails, and can be wheeled clear of the telescope when observing is to be undertaken. The shed for my 12½-inch has proper rails and wheels (it was built many years ago, when prices were very different from those of today), but angle-iron and pram wheels

Fig. 3.2. Run-off roof observatory for my 5-inch refractor.

will do just as well, provided that they are set in concrete so that they do not shift. I recommend making the shed in two parts, as shown here. If you have a single shed, it must have a door at one end, and this door must be either hinged or else removable. If it is hinged, it tends to flap; if removable, and you are trying to replace it in the dead of night with a strong wind blowing, it can act as a powerful sail. I can only add that the run-off shed for my 12½-inch telescope has stood since the end of the war and has been completely satisfactory; it is wooden, and completely waterproof, though I have known good run-offs made of hardboard covered with roofing felt. On the debit side, the observer is unprotected when he is working, which means that he is affected by any breeze and any stray light polution.

For a refractor, a run-off roof is suitable (Fig. 3.2). This is simple enough to make for anyone who, unlike myself, is good with his hands. With my 5-inch refractor, the roof runs back on rails which are supported at the end by two posts – actually the halves of an old flag-pole which I happened to find in my garden when I moved into my present house. The run-off roof design is not so suitable for a reflector, because it restricts the view of the sky.

Figure 3.3 shows the observatory which houses my 8½-inch Newtonian reflector, which has a German mounting. The base of the observatory is wooden, and the revolving upper section has glass windows, above which is the wooden roof – with a slit which can be opened from within. (Actually I do not recommend glass, but in my old house the only site for the observatory was the middle of the front lawn, and I wanted to make it look as decorative as possible.) The slit is asymmetrical, so that the telescope can be pointed vertically upward if need be. The main problem is making the

Fig. 3.3. Observatory housing my 8½-inch reflector at Selsey. The upper part of the building is rotatable.

ring upon which the upper section revolves. Obviously it has to be perfectly circular, as otherwise it will stick.

The same idea is used for the observatory containing my 15-inch, fork-mounted Newtonian (Fig. 3.3), though here there is no glass, and the whole construction is of metal, so that it looks rather like an oil-drum. The slit in the vertical part of the wall is simply hinged, and the slit in the roof is in two parts, which can be lifted up as shown in the diagram. Again I have had no real problems, though I admit that it takes some muscle to push the upper section around.

Yet another method is to make the whole observatory rotatable (Fig. 3.4). This 'shed' – actually, a converted garden shed – houses the 8-inch reflector belonging to Selsey amateur Reg Spry. It has the advantage of a door which can be entered without the need for the observer to stoop down, and it works very well, though I doubt whether the design would be practicable for any larger telescope simply because it would be too heavy to rotate manually.

Of course, there are many varieties of design, including the classic dome; all sorts of materials may be used, ranging from wood to plastic, metal or even roofing felt over a skeleton framework. (One thing which I do not favour is a slit which has to be taken out entirely every time the telescope is to be used; this can be a problem, particularly late at night.) I do not think that I need say more here, because the would-be builder will certainly be able to think things out for himself, and will design his observatory according to his needs, but I must give one more word of warning, because there is a trap into which I once nearly fell. When I moved house from Armagh, Northern Ireland to Selsey, Sussex, in 1968, I had the observatory for my 8½-inch set up in the

Fig. 3.4. Fully-rotatable observatory with an 8-inch reflector made and used by Reg Spry at Selsey.

garden. After the contract for sale had been signed, the purchaser claimed the observatory as being part of the deal. Frankly, I did not wait to argue; within twelve hours the observatory had been dismantled and removed. In fact there was no real danger, because the building stood freely upon its concrete platform and was not fastened down, but if it had been a permanent fixture I would have been in trouble – so beware!

What else does one need? For a reflector the eyepiece can often be high up, so that observing steps are essential; make them solid and broad, because, remember, you will be using them in total darkness for much of the time. Always have a torch handy, preferably with a red bulb, because if you switch on a powerful light you will ruin your night vision for at least an hour. For recording observations, have a pad fixed on to a board which has a dim red light attached. For some types of observation – variable stars, for instance – it is probably just as good to speak into a portable tape-recorder. Make sure that you have shelves or cupboards for your eyepieces and notebooks, and also make sure that you know exactly where they are. Obviously the observatory cannot be heated, and it may become hot during the daytime, so remember to open the slit – probably the door as well – some time before you plan to start work.

When making an observation, always note:

1. Date.
2. Time, always using GMT (Greenwich Mean Time) and eschewing horrible contrivances such as British Summer Time.

3. Type and aperture of telescope.
4. Magnification.
5. Seeing conditions, generally using the Antoniadi scale which is named after the Greek astronomer E.M. Antoniadi, who invented it.

If any of these pieces of data are omitted, the observation promptly loses most of its value. The Antoniadi scale is given in Appendix 6.

Always record your observations as soon as they have been made. The temptation to leave them in rough form, and transcribe them later, is a recipe for disaster. Personally, I have separate observing books for each subject — one for the Moon, one for Mars, one for variable stars and so on — though I know some observers who prefer to have one book only. If you are working with a team, you will have to send your observations away for analysis. Never trust originals to the post; always send copies. My observing books, which go back to 1934, never leave my study. Never throw an observation away, because you never know when it may suddenly become important.

Finally, and perhaps most important of all, avoid 'seeing' what you think you ought to see. I know, only too well, that this is a difficult matter, and that unconscious prejudice is hard to avoid, but always take the greatest possible care. A faulty observation is not only useless, but may be actually misleading.

I can cite a typical case of this. Many years ago I was observing Saturn, using my $12\frac{1}{2}$-inch reflector. There had been a report of an unusual white spot near the planet's equator, and I was anxious to confirm it. I worked out where I thought it ought to be on the disk, and then went to the telescope. Conditions were not good, and I estimated them as about 3 on the Antoniadi scale, but when I first looked I felt fairly confident that the white spot was in view, and I duly recorded it. An hour or so later the seeing improved, and when I made another observation the white spot was absent. This was not surprising, because I had made a mistake in my calculations, and the spot was actually on the side of the planet which was turned away from the Earth. I had been influenced by what I had expected to see, and the episode taught me a lesson which I have never forgotten.

4

The Solar System

The Solar System may be insignificant in the Universe as a whole, but it is of supreme importance to us, simply because we happen to live inside it. Also, the Sun, Moon and planets look very imposing in our skies; it is often difficult to realize that the Sun is a very average star, that the Moon is a junior member of the Solar System, and that the planets – even the most brilliant of them – are tiny compared with the stars.

Even the furthest-known member of the planetary system lies within 5000 million miles of the Sun, and this is not much as against the 24 million million miles separating us from the nearest of the night-time stars. Of course, nobody can really visualize these distances; we simply have to accept them.

For observations of most of the members of the Solar System, a telescope is essential. Meteors can be studied with the naked eye; binoculars will show a vast amount of detail on the surface of the Moon, but it is useless to pretend that even powerful binoculars can give good views of the planets. But before going into detail, it will help to look at a plan of the Solar System to see how the orbits are arranged.

It is obvious at first glance that the System is divided into two well-defined parts (Fig. 4.1). First we have four small and comparatively close-in worlds: Mercury, Venus, the Earth and Mars. Then comes a wide gap, in which move the thousands of asteroids or minor planets; beyond lie the four giants – Jupiter, Saturn, Uranus and Neptune – plus the remarkable little Pluto, which seems to be a maverick and which may not be worthy of true planetary status.

The individual movements of the bright planets have been known since very early times. The ancients knew that the planets keep strictly to a certain region of the sky which is called the Zodiac. This is because the orbits of the planets lie in much the same plane, so that if you draw a plan of the Solar System on a piece of flat paper you are not badly wrong. Consequently, the planets can be seen only in certain directions, and the same applies to the Sun and Moon. The Sun's apparent yearly path against the stars is known as the 'ecliptic'. (Of course the Sun and the stars cannot be seen at the same time under normal circumstances, because the sky is too bright.)

Early astronomers grouped the stars into constellations, and there are twelve constellations in the Zodiacal band, which stretches right round the sky. The first of these groups is Aries, the Ram. It contains no very bright stars, but in ancient times Aries was the constellation in which the ecliptic cut the celestial equator, i.e. the

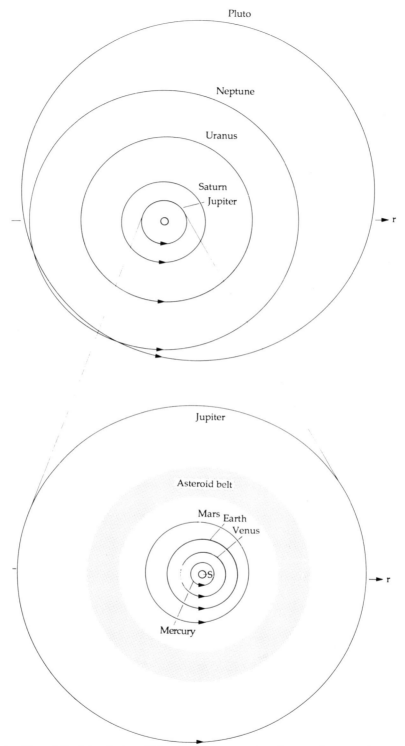

Fig. 4.1. Plan of the Solar System. (It has to be in two sections, because the outer planets are so remote as compared with the inner planets.)

projection of the Earth's equator on to the celestial sphere. Actually, the point of intersection, known as the vernal equinox, has shifted since then, because of the wandering of the polar point, and has now passed into the neighbouring constellation of Pisces, the Fishes; but it is still always called the First Point of Aries.

Since the planets are never far from the ecliptic, they are easy to recognize. In any case, Mars (when at its brightest) and Jupiter are too brilliant to be mistaken for stars, and Venus and Mercury, which are closer to the Sun than we are, have their own way of behaving. Only Saturn, and Mars when at its faintest, can cause any real confusion.

The first astronomer to give a proper description of the way in which the planets move was Johannes Kepler. Between 1609 and 1618 he published his three famous Laws of Planetary Motion:

Law 1. The planets move in ellipses. The Sun lies at one focus of the ellipse, while the other focus is empty.

Law 2. The radius vector (i.e. the line joining the centre of the planet to the centre of the Sun) sweeps out equal areas of space in equal times.

Law 3. The square of the sidereal period, or time taken to complete one orbit, is proportional to the cube of the planet's mean distance from the Sun.

These may sound rather complicated, but really they are quite straightforward. Law 1 is obvious enough; the only point to bear in mind is that although the orbits of the planets are ellipses, they are of low eccentricity, and are not very different from circles (except in the case of Pluto, which seems to be in a class of its own). It is the other two Laws which sometimes make beginners wrinkle their brows.

Law No. 2 is illustrated in Figure 4.2. The figure is not to scale, and the orbit or our supposed planet P is much more eccentric than is actually the case with any known planet in the Solar System (it is more like that of a comet), but one has to make the diagram distorted in order to make it clear. S stands for the Sun; P, P1, P2 and P3 represent the planet in various positions in its orbit.

Assume that the planet moves from P to P1 in the same time that it takes to go from P2 to P3. Then the area bounded by PSP1 must be equal to the area bounded by P2SP3. Since the dotted area is 'longer and thinner', it is evident that the planet is moving quickest when it is closest to the Sun. In other words, 'the nearer, the faster'. For example, Mercury has an orbit which is appreciably eccentric, so that at its closest to the Sun ('perihelion') it is only $28\frac{1}{2}$ million miles out, as compared with $43\frac{1}{2}$ million miles at its furthest point ('aphelion'). The orbital speed varies from $36\frac{1}{2}$ miles per second at perihelion to only 24 miles per second at aphelion. The Earth, moving in a more circular orbit at the greater distance of 93 million miles, has an average rate of a mere $18\frac{1}{2}$ miles per second, or 66 000 miles per hour.

The Third Law is very significant. The sidereal period — sometimes called the periodic time — has a definite link with the actual distance from the Sun. If we know one, we can find the other.

The Earth's sidereal period is $365\frac{1}{4}$ days. We can find out the periods of the other planets by simple observation; they range from 88 days for Mercury out to almost 248 years for Pluto. From Kepler's law, we can then draw up a complete scale model of the Solar System in terms of the 'astronomical unit', the distance between the Earth and the Sun.

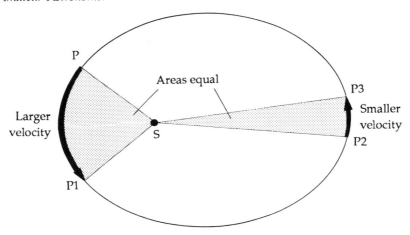

Fig. 4.2. Kepler's Second Law. The area PSP¹ must be equal to the area P²P³, assuming that the planet takes the same time to move from P to P1 as it does from P2 to P3.

To turn these relative distances into actual miles, all that is needed is one precise measurement. If, for instance, we can obtain an accurate value for the distance of Venus, we can work out the length of the astronomical unit by comparatively simple mathematics. The modern method is to bounce radar waves off Venus, and see how long they take to travel there and back, remembering that radar waves travel at the same speed as light (186 000 miles per second). We now know that the astronomical unit has a length of 92 957 000 miles.

The only natural body which genuinely moves round the Earth is, of course, our familiar Moon. Everyone is familiar with its monthly phases, from new to full and back again to new, but not everyone is sure how they are caused.

The Moon, like the planets, is a non-luminous globe, shining only by reflected sunlight. As the Sun can light up only one-half of the Moon at any time, the other half must be dark. In Figure 4.3, the Moon is shown in eight positions in its monthly journey – M1 to M8. At M1, the dark side is turned toward us; since this does not shine, the Moon is invisible, or *new*. As it moves toward M2, a little of the bright hemisphere starts to turn in our direction, and we see the Moon as a crescent; by the time M3 is reached, the phase has grown to half, or 'dichotomy', a position which is known, rather misleadingly, as First Quarter. Between M3 and M5 the phase is 'gibbous', between half and full, and by the time M5 is reached the Moon shows the whole of its day hemisphere. After full, the phase wanes once more, to half at M7 (Last Quarter) and then crescent at M8, until M1 is reached at the next new moon. The Moon takes 27.3 days to complete one orbit; but since both Earth and Moon are revolving together round the Sun, the interval between one new moon and the next is 29½ days.

If the lining-up at new moon is exact, the Moon blots out the Sun briefly and produces a solar eclipse. If the alignment occurs at full moon, the Moon passes into the shadow cast by the Earth, and we see a lunar eclipse. Eclipses do not happen every

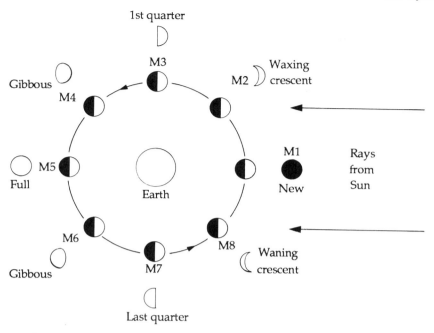

Fig. 4.3. Phases of the Moon.

month, because the Moon's orbit is appreciably tilted. I will have more to say about them later.

When telescopes were invented, it was soon found that Mercury and Venus also show phases. Figure 4.4 shows the behaviour of Venus. E represents the Earth, which we may assume to be stationary (really, of course, it is moving round the Sun, but let us make things as simple as possible). S is the Sun, and V1–V4 represent Venus in four different positions. Since Venus is closer to the Sun than we are, and moves more quickly, it completes one circuit in only 224.7 terrestrial days.

At V1 the Earth, Venus and the Sun are in a line, with Venus in the middle. The planet's dark side is then turned toward us, and Venus is new; this is termed 'inferior conjunction'. Occasionally the alignment is perfect, and Venus can be seen as a black spot against the solar disk; but since Venus too has a slightly tilted orbit, these transits are rare. The last was that of 1882, and the next will not occur until 2004.

As Venus moves on toward V2, we start to see the sunlit side. The planet appears in the morning sky as a slender crescent, becoming brighter and brighter as it draws away from the region of the Sun; the crescent can be seen with almost any telescope, or even with good binoculars. At V2 the three bodies make up a right-angled triangle, so that Venus appears as a half-disk (dichotomy). It then rises some hours before the Sun, and is a splendid sight in the east before dawn. At V2 it is said to be at Western or Morning Elongation.

As it travels on toward V3, Venus changes from a half into a gibbous form, and draws back toward the direction of the Sun, so that it becomes less and less prominent.

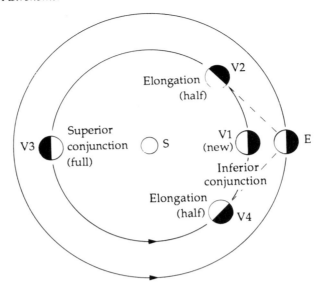

Fig. 4.4. Phases of Venus (or Mercury).

By the time it has reached V3, at 'superior conjunction', it is full, but since it lies almost behind the Sun it is not easy to find even with a telescope equipped with setting circles.

After passing superior conjunction, Venus reappears low down in the evening sky, shrinking gradually to a half as its angular distance from the Sun increases. It reaches Eastern or Evening Elongation at V4, and is then half-phase once more, after which it narrows to a crescent before returning to inferior conjunction at V1.

The synodic period of Venus, i.e. the interval between one inferior conjunction and the next, is 584 days, though this may vary by as much as four days either way. The interval between V4 and V2 is about 144 days, while 440 days are needed for the much longer journey between V2 and V4.

Venus is closest to the Earth at inferior conjunction. The distance is then no more than 24 million miles, about 100 times that of the Moon, but to all intents and purposes we cannot see it at all. When full, Venus is a long way away. All in all, it is a most infuriating object from the viewpoint of the observer!

Mercury behaves in the same manner as Venus, but since it is smaller, as well as being closer to the Sun, it is even more of a problem. It is never conspicuous with the naked eye, and there must be many people who have never seen it at all, though it is easy enough to find when at its best. Actually it can be more brilliant than any star; the trouble is that we never see it against even a reasonably dark background.

The other planets lie beyond the Earth in the Solar System, so that they cannot appear as crescents or halves, and can never pass through inferior conjunction, though Mars in particular can show up as decidedly gibbous.

In Figure 4.5, we see the orbits of the Earth and Mars. When the Earth is at E1 and Mars at M1, the Sun, Earth and Mars are lined up, so that Mars is at 'opposition', and is

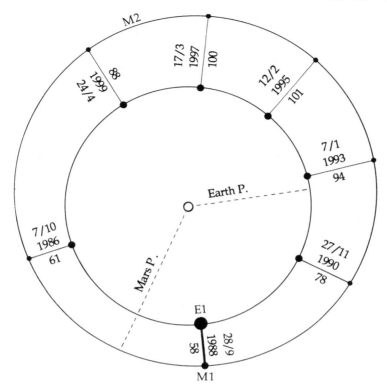

Fig. 4.5. Oppositions of Mars, 1968–99. The opposition of 1988 was the most favourable.

directly opposite to the Sun in the sky. A year later the Earth will have completed one revolution, and will be back at E1; but Mars, moving more slowly in a larger orbit, will not have had time to return to M1. It will have travelled only as far as M2, and will lie on the far side of the Sun, badly placed for observation. The Earth has to catch it up, with Mars moving onward all the time, and on average 780 days elapse before the three are lined up again. The 780-day interval between successive oppositions is therefore the synodic period of Mars, though again there are some variations, because the orbit of Mars is more eccentric than ours. When at quadrature, i.e. at right angles to the Sun, Mars shows a phase like that of the Moon a few days from full.

The other planets are much further out, and move so much more slowly that the Earth takes less time to catch them up. Jupiter's synodic period is 399 days, while that of Pluto is only $366\frac{3}{4}$ days. After having completed one circuit of the Sun, the Earth has only to travel for an extra day-and-a-half to overtake sluggish Pluto.

Most of the planets have satellites. The Earth, of course, has one; Saturn has 17, Jupiter 16, Uranus 15, Neptune and Mars two each and Pluto one, so that only Mercury and Venus are solitary travellers. Some of the satellites are of considerable size, and one – Ganymede, in Jupiter's System – has a diameter greater than that of Mercury.

Fig. 4.6. The Leonid Meteors, as seen from Arizona on 17 November 1966. The rate was about 2300 per minute. The meteors are seen here as streaks against the starry background; it is easy to recognize Ursa Major.

The asteroids or minor planets swarm in the wide gap between the orbits of Mars and Jupiter. All are cosmical dwarfs; only one (Ceres) is as much as 500 miles in diameter, and only one (Vesta) is ever visible with the naked eye. Even with a telescope they look exactly like stars, and the only way to identify them is to watch them from night to night, so that they can be seen to shift slowly against the starry background. Some asteroids swing away from the main swarm, and may even come close to the Earth, but all these so-called 'Earth-grazers' are small and faint, only a few miles in diameter.

Next we must consider the comets, which are much less substantial. Most of them move in orbits which are much more eccentric than those of the planets, and because they too depend upon sunlight they can be bright only when they are reasonably close to us. In fact, all the comets of short period (that is to say, a few years) are dim. The only bright comet which comes back regularly is Halley's, whose path takes it from inside the orbit of the Earth out to well beyond Neptune; the period is, on average, 76 years – but even Halley's Comet was faint at its last return, in 1986. Now and again we see really brilliant comets, but these have periods of many centuries, so that we never know when or where to expect them.

A large comet is made up of an icy core or nucleus, surrounded by a head or 'coma' of dust and gas. There may also be a long tail or tails, stretching out for millions of miles. Yet even a bright comet is a flimsy thing, of very slight mass, and is much less important than it may look. Note, too, that a comet lies millions of miles away, and cannot be seen shifting obviously across the sky. If you see something which is moving perceptibly, it certainly cannot be a comet.

Meteors, or shooting-stars, are cometary débris. They move round the Sun, usually in the paths of their parent comets, but are too faint to be seen unless they come close enough to dash into the Earth's upper atmosphere, when they become hot by friction against the air-particles, and burn away in the streaks of radiance which we call shooting-stars, ending the journey to the ground in the form of fine dust (Fig. 4.6). It is easy to prove that air sets up resistance; all you have to do is cup your hand and swing it around. Small wonder that a sand grain-size meteor, moving at anything up to 45 miles per second, will become violently heated. Above a height of 120 miles or so, the atmosphere is too thin to set up appreciable resistance. Larger bodies may land without being burned away, and are then termed meteorites, but there is no real link between a meteorite and a meteor. Meteorites seem to come from the asteroid belt, and there may be no difference between a small asteroid and a large meteorite.

Such is the Solar System. It contains bodies of all kinds, from the vast, intensely luminous Sun down to tiny particles of interplanetary dust. And it is, of course, the only part of the universe which we can explore with our rockets.

Until after the end of the war, professional astronomers in general paid scant attention to the Moon and planets, which were widely regarded as rather dull and parochial. The result was that amateurs had the field more of less to themselves, and, for example, the best lunar maps were of amateur construction. The situation changed when the Space Age began, in 1957. Since then men have landed on the Moon, and all the planets out as far as Neptune have been studied by rocket vehicles. Is there, then, anything left for the amateur to do?

The answer is an emphatic 'Yes'. At the moment (1989) the only planet which is being surveyed by a space-probe is Venus; if anything dramatic happens on Jupiter, or a major dust-storm covers Mars, it may well be an amateur who first detects it. Few professional astronomers have either the time or the inclination to concentrate upon such observations. So do not be discouraged; despite the Apollo missions, the Mariners, the Vikings and the Voyagers. There is still a great deal that we do not know, and amateur astronomers can be of real help, quite apart from the sheer enjoyment of looking at the deserts of Mars or the glorious rings of Saturn.

5

The Sun

In most branches of observation, the astronomer's call is for 'More light!' Things are different in the case of the Sun. There is plenty of light to spare, but there is also a great deal of heat, and it is essential to take the greatest care.

If you look straight at the Sun, you will dazzle yourself. Use a telescope, or a pair of binoculars, and your fate will be much worse, as permanent damage to the eye is likely. It is true that there are various gadgets which can be used to cut down the solar emission to an acceptable level, but I have never used them myself, and neither will I ever do so. The main trap is that some telescopes (mainly refractors) are supplied with dark filters, which, it is claimed, can be fitted over the eyepiece for direct observation of the Sun. Unfortunately, no filter can give full protection, and it is always liable to shatter without warning, in which case the observer may not have time to move his eye away from the telescope. There is only one rule about looking direct at the Sun through any telescope: *Don't.*

I have often been accused of over-dramatism here, and certainly I have given the warning more times than I can count in books, lectures, broadcasts and television programmes, but I am quite unrepentant. I joined the British Astronomical Association in 1934, when I was at the tender age of 11. At my first meeting, I made the acquaintance of an old amateur astronomer, then well in his eighties. He told me that he had been blind in one eye ever since he was 14. He had been viewing the Sun through a 3-inch refractor equipped with a solar filter – and the filter suddenly broke.

Having made this point, I must pause to say something about the Sun itself.

It is, as we have seen, a very ordinary star, but it is much larger than our world; its diameter is 865 000 miles, 109 times that of the Earth. But though its vast globe could contain over a million bodies the volume of the Earth, it is not as massive as a million Earths; the mass ratio is 332 000 to 1. Clearly, then, the Sun's mean density is less than that of the Earth, and is only 1.4 times greater than that of water.

Of course, this is not the uniform density throughout the globe. Density increases with depth. Near the centre of the Sun the material is denser than steel, even though it is still technically a gas.

The gravitational force acting upon a man standing on the surface of a globe depends upon two factors: the mass and the size. Taking the Earth's surface gravity as unity, the surface gravity of another body can be found by dividing the mass by the

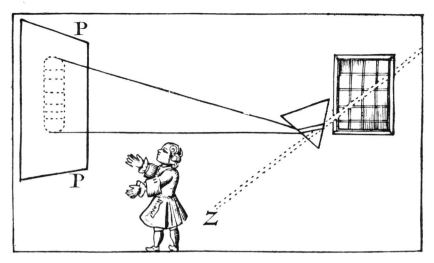

Fig. 5.1. Newton's experiment with sunlight.

square of the radius. For the Sun, these figures are respectively 332 000 and 109, so that the surface gravity is 332 000 divided by (109)², or 28. If you weigh 14 stone on Earth, you would weigh 2¼ tons if you went to the surface of the Sun, which would be distinctly uncomfortable. Not that it is likely to be put to the test; the surface is not solid, and the temperature is not far short of 6000°C.

The Sun is made up entirely of gas, though near the core this gas is under tremendous pressure – at least 1000 million atmospheres – and behaves in a decidedly un-gaslike manner judged by our normal standards. And this brings me on to a very brief discussion of how the Sun produces its energy.

It is not 'burning' in the usual sense of the term. Its energy is produced by nuclear reactions taking place near the core. The Sun contains a vast amount of hydrogen, which is the lightest of the elements and also the most plentiful (in the Universe as a whole, atoms of hydrogen far outnumber the atoms of all other elements put together). Deep inside the Sun, where the temperature is at least 14 000 000°C and possibly rather more, the nuclei of hydrogen atoms are combining to make nuclei of the second lightest element, helium. It takes four hydrogen nuclei to make one helium nucleus; every time this happens, a little energy is released and a little mass is lost. It is this energy which keeps the Sun shining. The mass-loss amounts to 4 000 000 tons per second, but please do not be alarmed; the Sun is at least 5000 million years old, and nothing spectacular will happen to it for another 5000 million years or so. There is no immediate need to pack our bags and search for a safer dwelling-place.

Analysis of the Sun's visible surface is carried out by means of the spectroscope. The first man to attempt anything of the kind was Newton, in 1666. What he did was to cut a small hole in the shutter of his window, so that only a narrow beam of sunlight could pass through; the light was then passed through a glass prism, and the resulting rainbow spectrum was spread out on to a screen (Fig. 5.1). This pioneer experiment can be improved by using a slit instead of a hole, and using a lens to bring the colours

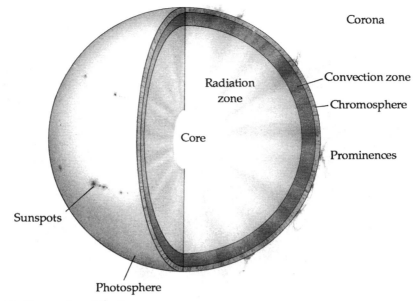

Fig. 5.2. Cross-section of the Sun.

to sharp focus. Newton never took his experiments much further, and the next real advance was delayed until the early nineteenth century. A British scientist, W.H. Wollaston, found that the Sun's spectrum was crossed by dark lines; but he thought that these lines were nothing more than the boundaries between the various colours, so that he missed the chance of making a discovery of fundamental importance.

Twelve years later, in 1814, the dark lines were seen again by the brilliant young German optician, Josef Fraunhofer, and Fraunhofer — unlike Wollaston — realized that they were of immense significance. They were always of the same intensity, and in the same places in the spectrum, so that it was possible to map hundreds of them. We still often call them the Fraunhofer Lines (Fig. 5.3).

All the matter in the Universe, whether in the Earth, the Sun or a remote galaxy, is made up of different combinations of a relatively small number of fundamental elements. There are 92 familiar elements, hydrogen being the lightest and uranium the heaviest; since they form a complete series, there is no chance of our having missed one. (Extra elements have been made artificially in recent years, plutonium being a famous — or perhaps notorious — example; but all these 'lead on' from the heavy end of the sequence, and probably do not occur naturally.)

When observed with a spectroscope, the bright surface of the Sun, which we call the photosphere, gives the bright rainbow first seen by Newton. Above this is a layer of more rarefied incandescent gas, extending upward for thousands of miles. On its own, this layer would give not a rainbow, but a number of isolated bright lines. However, seen against the rainbow background, the lines are 'reversed', and appear dark. The upper layer is called the chromosphere. A cross-section of the Sun is shown in Figure 5.2.

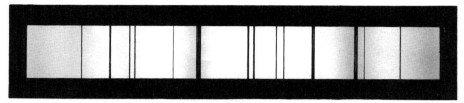

Fig. 5.3. The Fraunhofer lines in the solar spectrum.

The vital point is that the positions and intensities of the lines are unchanged. The spectra of the various elements can be studied in the laboratory, so that all we have to do is to compare the laboratory lines with the Fraunhofer lines in the spectrum of the Sun. For example, incandescent sodium gives two bright yellow lines (as well as many others). In the solar spectrum, there are two dark lines in the yellow part of the rainbow, corresponding exactly in position and intensity. The conclusion is inescapable: there is sodium in the Sun – and by now we have identified over 70 of the known elements. One of them, helium, was actually discovered in the Sun before it was identified on Earth.

The chromosphere, made up largely of hydrogen, cannot normally be seen with the naked eye or with ordinary telescopes, because it is so completely drowned by the glare of the photosphere; only during the fleeting moments of a total solar eclipse does it show up. We can then also see the Sun's outer atmosphere, the lovely corona, which extends for many millions of miles.

The most interesting features of the chromosphere are the prominences, which were once – misleadingly – known as Red Flames (Fig. 5.4). (It was only a century and a half ago that astronomers finally realized that they belong to the Sun rather than the Moon!) With special filters, or with spectroscopic equipment, they can be observed at any time, without waiting for an eclipse; but the equipment needed is expensive and complicated, so that the average amateur is limited to the features of the photosphere. Of these, the most obvious are, of course, sunspots.

Sunspots can be seen with any telescope (not by direct viewing, please; I will describe the safe method in a few moments). Basically, they look like dark patches on the brilliant surface (Fig. 5.5). They are not genuinely dark, and if they could be seen shining on their own the surface brightness would be greater than that of an arc-lamp; but they are around 2000°C cooler than the surrounding photosphere, so that they appear blackish by contrast.

A large spot is made up of a dark central portion (umbra) and a lighter surrounding area (penumbra) (Fig. 5.6). Several umbræ may be contained in one mass of penumbra; sometimes the shape of the complete spot is circular, sometimes the outline is complex and irregular. Small spots may be made up entirely of umbra, while in major spot-groups the penumbra is widely scattered.

Every spot-group has its own characteristics, but in general a typical 'two-spot' group begins as two tiny pores. The pores develop slowly into proper spots, growing and separating in longitude. Within a couple of weeks the group has reached its

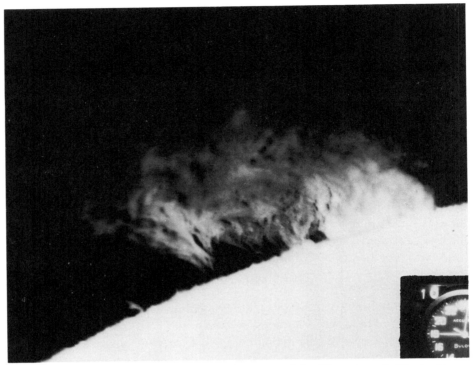

Fig. 5.4. Solar prominences. Photographs from the Big Bear Solar Observatory, California: 26 May 1978.

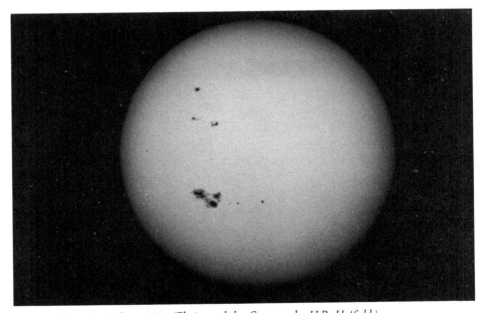

Fig. 5.5. Sunspots: 30 June 1988. (Photograph by Commander H.R. Hatfield.)

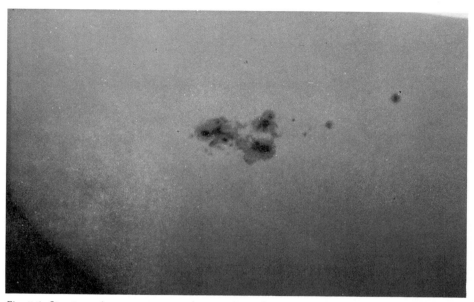

Fig. 5.6. Structure of a sunspot group, showing the umbra and the irregular penumbra: 30 June 1988. (Photograph by Commander H.R. Hatfield.) This group is also shown, to a smaller scale, in Fig. 5.5.)

maximum length, with a fairly regular leading spot and a less regular follower; there may also be many other small spots and clusters. After the group has reached its peak, a slow decline sets in. The leader is usually the last survivor, though not always. Roughly 75 per cent of groups follow this general pattern, but others do not conform, and single spots are also very common.

Small spots do not last for long, but large groups are much more persistent (Fig. 5.7). One group, which appeared in June 1943, did not finally vanish until the following December, a total period of around 200 days, but it was not under observation for the whole of that time. Since the Sun rotates on its axis, taking rather less than a month to do so, a spot-group is carried slowly across the disk from day-to-day; it will eventually be carried over the limb, and will not be seen again for about a fortnight, when it will reappear at the opposite limb – if, of course, it still exists.

We know how sunspots behave, but we have to admit that we are still not absolutely certain why they occur. Certainly they are the sites of powerful magnetic fields, and the theory proposed by H. Babcock in 1961 may be fairly near the truth. It is assumed that the Sun's magnetic lines of force run from one magnetic pole to the other, below the bright surface. The Sun does not rotate as a solid body would do; the rotation is 'differential', so that the equatorial regions spin the fastest and the polar regions the slowest. Over a period of years, the lines are distorted and drawn out into loops; eventually a loop of magnetic energy erupts through the surface, cooling it and producing the leading and following spots of a typical group. After about 11 years,

Fig. 5.7. Drawings of a sunspot group; 1989 January 26, 27 and 29 (January 28 was cloudy). I made these drawings with a power of × 100 on my 5-inch refractor, stopped down to 4 inches. Note the changes from day to day. By January 29 the group was approaching the limb of the Sun.

the lines have become so tangled that they break, and the Sun 'snaps back' to its original state. Certainly there is a well-defined solar cycle, which, incidentally, was first noted by an amateur astronomer, H. Schwabe, in the mid-nineteenth century. Every 11 years the Sun is at its most active, with many spots and spot-groups; things then quieten down, until there may be many consecutive days, or even weeks, with no spots at all. After minimum, activity builds up once more to the next maximum. Thus there was a maximum in 1980, and during 1985–86 the disk was quiet; the next maximum may be expected around 1991.

I say 'expected', because one never quite knows. The solar cycle is by no means perfectly regular, and the interval between successive maxima has been known to be as long as 17 years or as short as 7.3 years (Fig. 5.8). Moreover, there was one period – often called the Maunder Minimum – which lasted from 1645 to 1715, and during which spots seem to have been very rare indeed, so that the normal cycle was suspended. We do not pretend to know why, and this shows us yet again that our knowledge of our 'daytime star' is still far from complete.

Note too that not all maxima are equally energetic. Those of 1947–48 and 1957–58 were particularly violent, and the great group of April 1947 was the largest ever seen; at the peak of its development it extended over an area of over 7000 million square miles, as I well remember. Nothing on the same scale was seen during the last maximum, that of 1980.

The Sun was relatively 'quiet' during the mid-1980s, but activity then started to build up once more, and there were some large spot-groups in 1987 and 1988. Then, in March 1989, appeared a huge spot-group, which was associated with a brilliant display of aurora – the best for many years. This particular spot-group survived for several rotations of the Sun.

During the early part of a cycle, the spots tend to appear some way from the Sun's equator, but as the cycle progresses the spots invade lower and lower latitudes. As the cycle draws to its end, and its groups die away, small spots of the new cycle start to appear at high latitudes once more. Near minimum, then, there are two areas subject to spots; low latitudes, with the last spots of the dying cycle, and higher latitudes, with the first spots of the new cycle. This behaviour follows what is called Spörer's Law, first announced in 1879 by the German astronomer of that name. Note, however, that spots are never found either at the actual equator or at the poles.

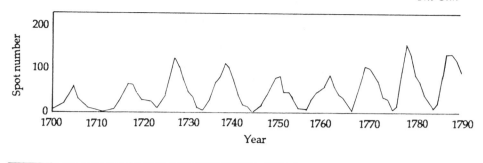

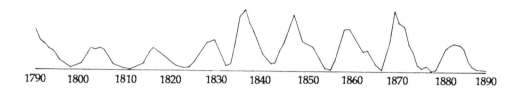

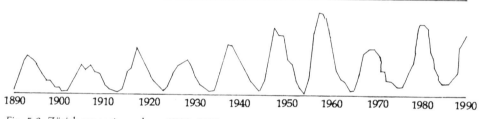

Fig. 5.8. Zürich sunspot numbers, 1700–1989.

Spots are often associated with bright irregular patches known as faculæ, from the Latin word meaning 'torches'. Faculæ lie well above the photosphere, and may be classed as luminous 'clouds' in the upper regions. They often appear in regions where a spot-group is about to break out, and may persist for some time after a group has disappeared. If you see obvious faculæ on the Sun's following limb, you may be fairly confident that a spot-group will shortly come into view from the far side.

Solar flares are brilliant, short-lived outbreaks in the chromosphere in the region of an active spot; large amounts of energy are released, and they spread rapidly through large areas of the chromosphere horizontally. They produce marked effects on the Earth, notably 'magnetic storms' (variations in the compass needle), interference with radio transmission, and the production of auroræ or polar lights, about which I will have more to say later. Unfortunately they are very seldom visible in integrated light, and have to be studied with spectroscopic equipment. I admit that I have never seen one with an ordinary telescope, though they have been recorded now and then (initially by two amateurs, Carrington and Hodgson, in 1859).

Even on the unspotted parts of the Sun, ceaseless activity is going on. The photosphere is never calm; it is covered with 'granulations', with each granule having an average width of around 500 miles. A granule may last for eight minutes or so, and it has been estimated that the photosphere includes about four million granules at any one time. They are upcurrents, and the general situation has been likened to the boiling of a liquid, bearing in mind that the photosphere is purely gaseous.

There is much to be said about sunspots and their behaviour; but this is not a book about theoretical astronomy, so let me turn now to what the would-be solar observer should do, assuming that he is equipped with nothing more elaborate than an ordinary telescope. Always remember the dangers. If the Sun is seen shining through a thick layer of mist or haze, it may appear reassuringly dim and gentle; the temptation then is to use a telescope for a direct view, either with or without a suncap. Anyone who succumbs may well suffer disastrous eye damage. I do not even recommend looking at the Sun with no optical aid at all. True, there is no harm in a fairly brief observation provided that a piece of very dark glass is used, and some sunspots are large enough to be seen with the naked eye; but again, it is something which I personally never do.

Fortunately there is an easy alternative – projection. First turn the telescope in the direction of the Sun, without putting your eye anywhere near the eyepiece. Then hold a white card a few inches away from the end of the tube and move the telescope gently (if necessary) until the image of the Sun appears, after which the disk can be brought to a sharp focus by adjusting the positions of the card and the rack of the eyepiece. Any spots and faculæ which happen to be present will be obvious at a glance. A low power is advisable for a whole-disk view – I have found that for my 3-inch f/12 refractor, × 72 gives good results – though the magnification can be increased for drawing individual groups on a larger scale.

To make the drawings conveniently standard, it is best to draw a 6-inch circle on the card and then adjust the distance and focus until the image of the Sun exactly fits it (Fig. 5.10). If the telescope used is very small, a 4-inch circle may be more suitable.

It is not easy to hold the card steady, move the telescope so as to follow the Sun, and draw the spot-groups all at the same time. You would need several hands. The obvious solution is to fit an attachment to the telescope tube which will hold the card at the right distance from the eyepiece; there is nothing difficult about this if you are even reasonably good with your hands, and it can still be done even if the observer is as clumsy as I am. The main thing to avoid is upsetting the balance of the telescope tube.

My own procedure follows a set pattern. First, locate the Sun and examine the whole of the disk, paying special attention to the following limb and looking for any faculæ which may herald the appearance of a new group. If there are any large spots on view, make up your mind which are to be sketched in detail, and then mark in the positions of the main components on the 6-inch circle. When this has been done, you can do the more detailed drawing at leisure. Then add the usual notes – date, time, telescope, power and seeing conditions.

If you aim to be a more serious observer, there are extra points to bear in mind. You may, for example, want to work out the heliocentric latitudes and longitudes of the spot-groups, and there are three things to determine: *P* (the position angle of the north

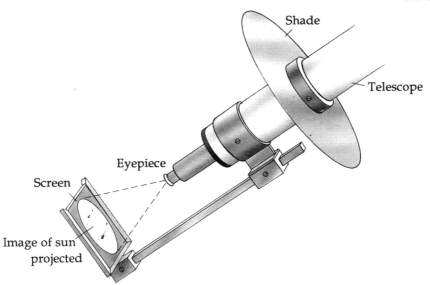

Fig. 5.9. *Projecting the Sun's image. This is the only safe way to observe sunspots; if you look direct, even with a dark filter, you will almost certainly ruin your eyesight. The telescope is used to send the Sun's image on to a white card or screen held or fixed behind the eyepiece; the shade is to cast a shadow over the screen.*

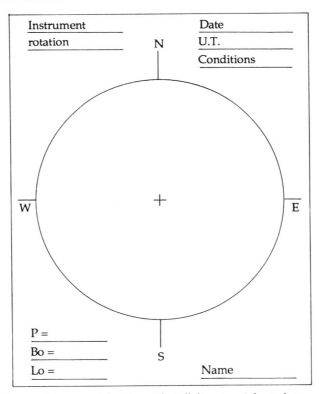

Fig. 5.10. *Blank for recording sunspot drawings. The full diameter of the circle is usually 6 inches.*

59

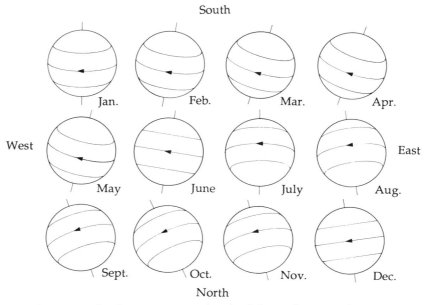

Fig. 5.11. Apparent tracks of a sunspot across the solar disk over the course of a year.

Fig. 5.12. The Wilson effect. When a spot is near the limb, its penumbra appears to be narrowest in the direction toward the centre of the disk. (Spot-size is exaggerated here.)

end of the Sun's axis of rotation, given as $+$ if east and $-$ if west); B_0, the latitude of the centre of the Sun's disk ($+$ if toward us, $-$ if away from us), and L_0 (the longitude of the centre of the disk, or of the central meridian). Tables of these three values for noon, at four-day intervals, are given in publications such as the *Handbook of the British Astronomical Association*, and intermediate values can easily be worked out. To determine L_0 you need a table, since the longitude decreases at the rate of 13.2 hours per day. I have given further details in Appendix 9.

To orient your drawing, mark a line across a diameter of your 6-inch circle to give the east–west direction. Then allow a spot to drift straight along the line, and draw a second line at right-angles to it. Only when this has been done should you put in the positions of the groups.

It is fascinating to follow the spots from day-to-day, and see how they are shifted across the disk because of the Sun's rotation. Except during early June and early December, you will find that the spots describe a curve, because of the apparent shift in position of the solar axis (Fig. 5.11). The position of the pole for any date can be looked up, but the rough diagrams given here will be a help.

It is useful to note the number of active areas; any spot, however small, is regarded as a separate active area if it is more than 10 degrees away from its nearest neighbour. I also make daily notes of what is called the Wolf or Zürich number. This number (R) is given officially by the formula $R = k (10g + f)$, g being the number of groups and f the total number of individual spots. k is a constant depending on the experience and equipment of the observer, but for most purposes it can be taken as 1. Therefore, to obtain a useful value for the Zürich number, all you have to do is to count the number of groups, multiply by 10, and then add the number of spots you can see all told.

One interesting phenomenon is the so-called Wilson Effect (Fig. 5.12). If a spot has a sunken umbra, the 'preceding' penumbra will be foreshortened, and will appear narrow as the spot comes over the limb, while when the spot has crossed the central meridian and is approaching the opposite limb the 'front' penumbra will be the broader. In fact, with a circular spot which has a depressed umbra, the penumbra closest to the Sun's centre will always seem to be narrower than that on the opposite side of the spot. Most spots seem to be relatively shallow hollows a few hundreds of miles deep, but now and then there are unusual spots where the effect is actually reversed.

So far as telescopes are concerned, refractors are generally more suitable than reflectors; if you are not using a projection box, fit a screen over the eyepiece end of the tube to cast a shadow on the screen (Fig. 5.9). A 3-inch refractor is very suitable. I use my 5-inch, but I often stop down the object-glass. A 6-inch is definitely too large, as it will collect too much light, so it must nearly always be stopped down. With a reflector, it is wise to leave the mirror unsilvered, which naturally makes it more or less useless for any other kind of observation – though again this can be overcome by judicious stopping-down. I know several observers who use reflectors but have two mirrors – one silvered, for ordinary use, and the other unsilvered, specially for the Sun. (By 'silvered' I naturally mean coated with some highly reflective substance; aluminium, or more infrequently rhodium.)

Spots and faculæ can of course be photographed. Solar cameras can be obtained, but a simpler method is to project the image and then simply photograph the screen. There is bound to be some distortion, because you will find that you simply cannot get your camera in the right position, but at least you will be able to record something.

There are some amateurs who carry out really valuable research with more complicated equipment. One British observer, Commander Henry Hatfield, has actually built his house round an instrument known as a spectrohelioscope. But while it would be idle to pretend that the observer who contents himself with drawing spots and faculæ with the aid of a small telescope has much chance of making a spectacular discovery, particularly since daily disk photographs are taken at some professional observatories, the time spent will not be wasted. It is fascinating to watch the spots as they appear, drift, change and finally die away. But at the risk of sounding tedious, let me repeat my warning once more: take care. A cat may look directly at a king, but no telescopic observer would ever look directly at the Sun.

6

The Moon

The Moon is our faithful companion. On average it is a mere 239 000 miles away from us, and in Britain, at least, many motorists will find that they are driving cars which have covered a much greater distance than that (mine certainly has!). Though the Moon is much smaller than the Earth, with a diameter of only 2160 miles, it dominates the night sky for a large part of every month.

You can see definite markings with the naked eye – who has not heard of the Man in the Moon? Any telescope, or any good binoculars, will show a vast amount of detail. There are mountains, valleys and craters; the sight of a lunar landscape is something never to be forgotten, and the Moon will always be a favourite object for amateur observers (Figs. 6.1, 6.2). Moreover, amateurs have carried out very useful work in lunar charting, and it is fair to say that before the start of the Space Age, all the best Moon-maps were of amateur construction.

Things are different today, of course. The Moon is no longer inaccessible; it has been reached, and many of its outstanding puzzles have been solved, though others remain. Quite apart from the manned landings of the Apollo programme, there have been automatic probes which have flown round and round the Moon, sending back photographs of amazing detail and quality, so that by now we have very accurate charts of virtually the whole of the surface. And the Russians have sent unmanned vehicles to the Moon, brought them down gently, and then re-launched them, bringing them home together with samples of lunar material.

In astronomy, as in everything else, honesty is the best policy, and we have to admit that in most ways, though not all, the amateur lunar observer has completed his task. There is no longer any scientific value in, say, making a chart of a lunar crater with the help of a 6-inch or 12-inch telescope purely for cartographic purposes. It is still worth doing, for the pleasure it gives the observer, but if any researcher wants to look for fine details he will turn at once to a photograph taken from close range. Yet there remains some really valuable work for the amateur to do, and the last thing I want is to be discouraging – if only because I have always been more concerned with observation of the Moon than with any other branch of astronomy.

The Moon is usually called the Earth's satellite, but this may be misleading; it is too large to be an ordinary satellite. Remember, its diameter is more than one-quarter that of the Earth, and very probably it has always been a separate body – the old theory

Fig. 6.1. The Moon. (left) eastern hemisphere; (right) western hemisphere. South is to the top.

When these photographs were taken, libration was favourable for the eastern side, so that the Mare Crisium is well away from the limb.

that the Moon was once part of the Earth, and broke away, has fallen into disfavour even though it still has some supporters. In my view, it is best to regard the Earth–Moon system as a double planet. Rock samples brought back by the Apollo astronauts and the unmanned Russian probes confirm that the ages of the two worlds are about the same (4.6 to 4.7 thousand million years).*

It is not strictly true to say simply that the Moon moves round the Earth; more accurately, both bodies move round the barycentre, or centre of gravity of the Earth–Moon system (Fig. 6.2). However, the barycentre lies deep inside the Earth's globe, because the Earth is so much more massive than the Moon; the ratio is 81 to 1, showing that in addition to being much the smaller of the two the Moon is also much less dense. On average its globe is only 3.3 times denser than water, as against 5.5 for the Earth, so that presumably the Moon has a much smaller 'heavy core' – a fact borne out by the space mission results.

The Moon's low mass means that it has a low escape velocity: $1\frac{1}{2}$ miles per second, as against 7 miles per second for the Earth. Even if it once had an atmosphere, it could not retain it, and today the Moon is 'airless'. Go there, and you will find that the sky is black even in the daytime. There are no winds, no storms and no 'weather'. Since there is no atmospheric shield, the surface becomes fiercely hot in the daytime (at least in the region of the lunar equator), while the nights are bitterly cold; the surface materials are very poor at holding on to the daytime heat.

The Moon is a slow spinner. Its axial rotation period is 27.3 days, which is the same as the time taken for the Moon to complete one journey round the Earth (or, more properly speaking, round the barycentre). The effect of this captured or synchronous rotation is that the Moon keeps the same face turned toward us all the time. This sometimes causes confusion, but an easy experiment will show what is meant. Place a chair in the middle of the room, to represent the Earth, and assume that your head is the Moon. Your face stands for the familiar hemisphere, while the back of your neck represents the 'back' of the Moon. Now walk round the chair, keeping your nose turned chairward all the time. When you have completed one circuit, you will have turned round once, because you will have faced every wall of the room; your 'sidereal period' will have been equal to your 'axial rotation', and anyone sitting on the chair will not have seen your back hair at all. This is how the Moon behaves. From Earth, we can never see the far side of the Moon.

There are, however, some qualifications. Though the Moon spins on its axis at a constant speed, it has an orbit which is slightly eccentric, so that its velocity changes. When at its closest to the Earth (perigee) it moves faster than when it is furthest away (apogee). The axial spin and the position in orbit become periodically out of step, and the Moon seems to rock slowly to and fro (Fig. 6.3), allowing us to see a little way first round one mean limb and then round the other. On some nights the grey plain of the Mare Crisium will appear to be almost touching the limb, while on others it will be well clear, and extra features come into view. There is also a rocking in a north–south direction, and a slight rocking also because of the spin of the Earth itself. These various

---

* Note that I studiously avoid using the term 'billion'. The old British billion was a million million, while the more common American billion is a mere thousand million.

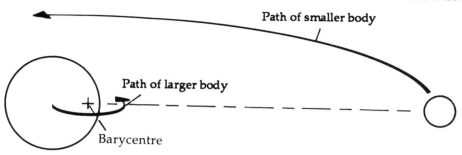

Fig. 6.2 Barycentre.

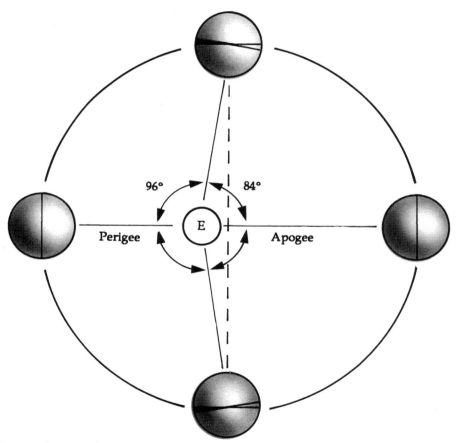

Fig. 6.3 Libration in longitude. The Moon spins on its axis at a constant rate; it moves in its orbit at a variable rate, travelling fastest near perigee, so that the position in orbit and the amount of rotation become periodically out of step, and the Moon seems to rock slightly to and fro. We can therefore see a little way round alternate mean limbs.

librations' mean that from Earth we can examine a grand total of 59 per cent of the total surface, though of course never more than 50 per cent at any one time; the remaining 41 per cent is always out of view, and until 1959, when the Russians sent their robot probe Lunik 3 on a 'round trip', we knew nothing definite about it. Incidentally, there is no mystery about this captured rotation; it has been caused by tidal friction over the ages, and most of the other planetary satellites in the Solar System behave in the same way. Thus Titan, the largest satellite of Saturn, has an orbital period of 15 days $22\frac{1}{2}$ hours; Titan's axial rotation period is exactly the same.

Remember, though, that while the Moon always keeps the same face turned toward the Earth, it does not keep the same face turned toward the Sun, so that day and night conditions are the same all over the Moon. The only real difference is that from the far side, the Earth will never be seen, so that the nights will be blacker because of the absence of Earthlight.

When the Moon is a crescent, the unlit side can often be seen shining faintly; this is called the Earthshine (Fig. 6.4), because it is due to light reflected on to the Moon from the Earth. It is often nicknamed 'the Old Moon in the Young Moon's arms', though it can also be seen with the waning moon a few mornings before returning to new. I must also mention the famous Moon Illusion, according to which the full moon looks much larger when low down than when high above the horizon. This *is* an illusion and nothing more, but I agree that it is very marked, and it was even mentioned by Ptolemy almost two thousand years ago!*

The first telescopic maps of the Moon were drawn not long after the invention of the telescope itself. Pride of place seems to go to an Englishman, Thomas Harriot (one-time tutor to Sir Walter Raleigh) who undoubtedly produced a chart in 1609, but the first really serious observer was the great Italian, Galileo, who in 1610 drew an outline map in which several features are clearly identifiable. Others followed, and in 1651 another Italian, Riccioli, introduced the system of naming craters after famous personalities – mainly, though not always, astronomers. Riccioli's system is still used, though of course it has been extended. Lunar mountains are often named after ranges on Earth, so that we find the Alps, the Apennines, the Caucasus and various others.

The real founders of what we call selenography were three Germans. Johann Hieronymus Schröter was the first, from 1778 to 1814; then, in 1838, his countrymen Beer and Mädler produced a map which was a masterpiece of careful, accurate observation. Later in the century came the first photographic atlases, and also some excellent visual charts. Before the Space Age, the largest of all Moon maps was completed by the Welsh amateur H. Percy Wilkins: it was 300 inches in diameter, and took him 40 years to complete. But then came the rockets, and professional workers turned their attention to the Moon. Robot probes obtained photographs of the far side; then, between August 1966 and August 1967, came the five unmanned American Orbiters, which were put into closed paths round the Moon and sent back thousands of high-grade pictures which at once made all previous maps obsolete. As

---

* With Professor Richard Gregory, I once conducted a somewhat hilarious television experiment to show that this is true. If you want details, you will find them in my book *TV Astronomer* (Harrap, 1987).

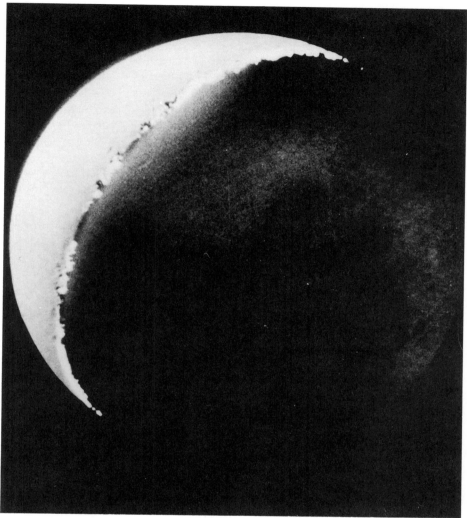

Fig. 6.4. The Earthshine, 1967 May 13. The bright crescent is necessarily over-exposed. (Photograph by Commander H.R. Hatfield.)

almost everyone will know, the first astronauts stepped on to the lunar surface in July 1969, and the Apollo programme went on until December 1972. Meanwhile, the Russians had sent their sample-and-return vehicles, and also two 'crawlers' which tracked around the Moon's surface, guided by their controllers in the USSR.

Where does all this leave the Earth-based amateur observer? It has been claimed that there is nothing left to do, but I strongly disagree. There is still a great deal that we do not know, and it is also a fact that the last lunar probe, Russia's Luna 24, dates back as far as August 1976, since when the Moon has been shamefully neglected. So let us take stock, and see in which ways the observer can contribute.

The most obvious markings on the Moon are the dark grey plains which are always known as seas (Latin, *maria*). They have never been water-filled, since we now know that there has been no water on the Moon throughout its long history, but they are old seas of lava, so that the names are not entirely misleading. Many of them, such as the vast Mare Imbrium or Sea of Showers, are roughly circular in outline; they are bordered by mountain chains, some of which are decidedly high. The Lunar Apennines, forming part of the boundary of the Mare Imbrium have peaks going up to around 15 000 feet; other sections of the boundary are formed by the Alps, cut through by a magnificent valley, and the Caucasus. Other seas notably the largest of all, the Oceanus Procellarum (Ocean of Storms) are less regular, while the Mare Frigoris (Sea of Cold) is more in the nature of a 'strip'. Most of the major maria are joined in one vast system, though a notable exception is the smaller, well-marked Mare Crisium (Sea of Crises), which is easily visible with the naked eye near the north eastern limb.* There are also some isolated but quite well-marked seas close to the limb, in what are termed the libration areas which are carried alternately in and out of view by the Moon's slight rocking; examples are the Mare Humboldtianum (Humboldt's Sea) and the Mare Marginis (Marginal Sea). Of special interest is the Mare Orientale (Eastern Sea). I have a fatherly interest in it, because I discovered it myself, observing together with H.P. Wilkins; we were using a modest telescope, and came across it when we were examining the libration regions, which were very favourably placed on that particular night. We completed a rough chart, and sent in a report, suggesting its name. What we did not then know, and had no means of knowing, was that it is in fact a huge ringed basin extending on to the Moon's far side. Ironically, the reversal of official 'east and west' on the Moon means that our *Eastern* Sea is now on the *western* limb!

This sort of observation – made soon after the war – is an indication of the sorts of discoveries which could then be made by amateurs, but we have to admit that this is no longer true, because space-research methods have taken over.

Some of the seas have bays leading out of them. The most beautiful is certainly the Sinus Iridum (Bay of Rainbows), off the Mare Imbrium. When the Sun is rising over it, the peaks to the limbward side are illuminated before the floor of the bay itself, producing the famous 'Jewelled Handle' effect. It lasts for only a few hours, but is worth waiting for.

Earth-type mountain chains are rare, but isolated peaks and clusters of hills are very common indeed; note for example the Harbinger Mountains, near the brilliant crater Aristarchus, and the strangely regular Straight Range, near the northern edge of the Mare Imbrium. Two bright isolated peaks, Pico and Piton, also lie on the Mare Imbrium, and are very prominent indeed.

The whole lunar scene is dominated by the walled circular formations which we

---

* Until 1966 it was officially recognized that west was to the left-hand side of the Moon, with south at the top. Then, at a congress of the International Astronomical Union, it was decreed that this should be reversed, so that Mare Crisium is now to the east and the dark-floored crater Grimaldi to the west. I opposed the change, but was heavily out-voted. In this book I have naturally bowed to the IAU decision; Mare Crisium to the east.

usually call craters. No part of the Moon is free from them. They abound in the bright uplands, and are also to be found on the maria, on the slopes of mountains and even on the crests of peaks. They break into each other and deform each other; some have massive, terraced walls and high central mountain structures, while others are low-walled and ruined, so that they come into the category of 'ghosts'. The largest craters – more appropriately termed walled plains – are well over 150 miles in diameter; the smallest are too tiny to be seen at all from Earth.

Though some of the craters are deep, with walls rising to well over 10 000 feet above their floors, they are not in the least like steep-sided mine-shafts. A typical crater has a mountain rampart which rises to only a modest height above the outer landscape, but much higher above the sunken interior, so that in profile a crater is more like a saucer than a well. The inner walls are often terraced. Erathosthenes, on the border between the Mare Imbrium and the Sinus Æstuum (Bay of Heats) is typical; it is 38 miles in diameter, with a massive central mountain structure which does not, however, attain the height of the outer rampart. With some craters, the central feature is a single peak; others have clusters of hills or central craters, and we also find examples of concentric craters, such as Vitello on the edge of the Mare Humorum (Sea of Humours) (Fig. 6.5). With some craters, such as Pitatus on the edge of the Mare Nubium (Sea of Clouds), the seaward wall has been greatly reduced by the mare lava, while with others, such as Le Monnier on the edge of the Mare Serenitatis (Sea of Serenity) (Fig. 6.6), the wall to the seaward side has been so levelled that it can barely be traced at all.

Some of the larger formations have relatively smooth floors, such as the 60-mile Plato, between the Mare Imbrium and the Mare Frigoris. Plato is dark-hued, but Ptolemæus, near the centre of the Moon's face as seen from Earth, has a brighter interior containing only a few craters and peaks – even though it is not nearly so smooth as it may look when seen through a telescope.

Plato, incidentally, is almost perfectly circular, but it looks oval, because it is not too far from the Moon's limb, and appears foreshortened. This effect is very marked for all formations except those near the centre of the disk, such as Ptolemæus. The Mare Crisium is a case in point. It looks elongated in a north–south direction, but is actually slightly larger when measured through its east–west diameter.

The extreme libration areas are difficult to map (as I know, inasmuch as I spent 30 years in charting them before the Space Age, and, I am glad to say, my results were used by the Russians to correlate the first far-side pictures with the features of the familiar side of the Moon). It is hard to tell the difference between a crater and a ridge, for example. Libration charting is still interesting, though we must admit that it is outmoded now.

How were the craters formed? Either they are due to internal action, or to meteoritic bombardment, or a mixture of both. No doubt craters of both kinds exist. I do not propose to discuss the matter here, if only because in supporting the view that the main walled plains and craters are of internal origin I know that I am in the minority; but it is a fact that the distribution of the craters is not random, and when one structure breaks into another it is the rule that it is the smaller crater which is the intruder.

Fig. 6.5. The Mare Humorum. The large crater at the lower northern end of the Mare is Gassendi; on the edge of the Mare, to the south, note the 'bay', Doppelmayer, with its levelled north wall and its central peak. (Photographed by Commander H.R. Hatfield on 3 March 1966, with his 12-inch reflector.)

Thebit, on the edge of the Mare Nubium, is typical of what I mean. Also the large structures, as well as the small ones, tend to form groups or chains; look for instance at the chain near the edge of the Mare Nectaris or Sea of Nectar (Theophilus, Cyrillus and Catharina), the Ptolemæus chain (Ptolemæus itself, Alphonsus and Arzachel) and the Walter chain in the uplands (Walter, Regiomontanus and Purbach).

There are many minor features of interest. The clefts or rills (often spelled 'rilles') look like cracks, and may extend well over 100 miles; a famous example is the Ariadæus Rill, running into the Mare Vaporum (Sea of Vapours) which is associated with a whole system. Inside the majestic crater Petavius, near the Mare Fœcunditatis (Sea of Fertility) a prominent rill runs from the central peak to the wall; some craters, such as Gassendi (on the edge of the Mare Humorum) (Fig. 6.5) and Alphonsus (in the Ptolemæus chain) have complicated interior rill-systems. On the other hand, some so-called rills prove to be made up of strings of small craters which have run together. Such is the conspicuous Rheita 'Valley', in the southern uplands, while the Hyginus Rill, near Ariadæus, is in part a crater-chain. The Alpine Valley is a huge gash cutting through the Alps, and extending from the crater Herodotus, near the brilliant Aristarchus, we see the winding valley which is named in honour of Johann Schröter, who discovered it almost two hundred years ago.

In the Mare Nubium, near Thebit, we find the Straight Wall, which is not perfectly straight and is certainly not a wall. It is a fault in the lunar surface, 80 miles long; the ground to the west drops by 800 feet, though the gradient is not steep (no more than about 40 degrees). Before full moon the Straight Wall shows up as a dark line, because of its shadow. After full, when the sunlight is striking the inclined face, the Wall looks bright.

Domes may be likened to gentle swellings in the lunar crust; many of them have summit craterlets, and some are riddled with fissures. And then, of course, there are the bright rays, extending out from a few major craters such as Tycho in the southern uplands and Copernicus in the Mare Nubium. The rays are surface deposits. They are barely visible under low illumination, but under high light they are so prominent that they dominate the scene entirely.

The first step for the would-be lunar observer is to learn his way around. It is hopeless to start any systematic programme until all the main features, and many of the minor ones, can be recognized on sight. Crater identification is not so difficult as might be thought, but there are some basic points to be borne in mind, and various traps to avoid. When a crater is near the terminator, or boundary between the daylit and night hemispheres of the Moon, it will have shadow inside it, and will be strikingly conspicuous; under higher illumination the shadows will shrink, and the same crater may be hard to find at all. Toward full moon the shadows almost disappear, and the craters are difficult to locate unless they have particularly bright walls (such as Aristarchus), particularly dark floors (such as Plato or Grimaldi) or are ray-centres (such as Tycho, Copernicus and Kepler). Full moon is not the best time to start observing – on the contrary it is the worst, particularly as the bright rays are so obtrusive. The most spectacular views are obtained during the crescent, half or moderately gibbous stages.

Fig. 6.6. Mare Serenitatis, as I photographed it with my 15-inch reflector. Linné is the white patch below the Mare centre: the upper craterlet on the Mare is Bessel.

I can cite a personal experience here. I first looked at the Moon through a telescope when I was a boy of eight; since I knew no better, I decided to make a start on the night of full moon. I looked up the position of the 92-mile crater Ptolemæus, arranged my newly-acquired 3-inch refractor, and tried to find my way around. Naturally, I failed to find Ptolemæus. When I looked again, at the next half-moon, the crater was filled with shadow, and I could not possibly miss it.

The method I then adopted – and which I still recommend – was to take an outline chart of the Moon and set out to make at least two sketches of every named feature. The procedure takes a long time, because a normal crater can be well-sketched only when there is some shadow inside it, and you have to make the most of limited opportunities (by Spode's Law, the skies are always cloudy at the wrong moment). By the time I had finished, the programme had taken me about two years. The sketches themselves were useless, as I knew they would be, but at least I had learned how to tell one crater from another.

It is a great mistake to make a drawing too small, or to attempt too large an area at any one time. Probably about 20 miles to the inch is a good scale. 'Finished' drawings look attractive, but an observer with no artistic gifts, such as myself, may be wiser to keep to line drawings. Accuracy is always the main objective. Always remember that a crater alters in appearance according to illumination, and it is necessary to be able to identify it under all possible conditions of lighting.

Structurally, the lunar surface is to all intents and purposes changeless. It must be 1000 million years or so since the last large crater was formed, and most are considerably older than that, though no doubt small impact craters are produced every now and then. One feature which has caused a great deal of discussion is Linné, on the Mare Serenitatis. Originally it was described as a deep craterlet, comparable with Bessel, the largest of the formations on the Mare Serenitatis, but since 1866 it has taken the form of a white patch with a small central craterlet. There have been suggestions that it has genuinely altered, but this seems most improbable, and the space-probe pictures show it as a perfectly normal object. Another case of reported change concerns the twin craters Messier and Messier A, on the Mare Fœcunditatis. The pioneer observers Beer and Mädler described them as being exactly alike; actually they are not, since A is the larger of the two and is less regular in outline. Yet they change in aspect according to the conditions of lighting, as you will soon see if you follow them through a complete lunation, and again there seems no evidence that any genuine alteration has taken place. Note, incidentally, the strange double ray extending west toward the mare border; it is often nicknamed the 'comet'. A small telescope will show it easily.

In recent years, however, there have been interesting observations of a different kind, involving temporary local obscurations. These are known as Transient Lunar Phenomena, or TLP – a term which I introduced long ago, and which has now come into general use.

The story of TLP goes back to the last century. Amateurs recorded the patches now and then, but their reports were regarded with some scepticism, because none could be photographically confirmed. This scepticism was unwise. Before about 1950 the only observers who were paying systematic attention to the Moon were amateurs – apart from a very few professionals, such as Edward Emerson Barnard and W.H. Pickering in America, who also recorded TLP now and then. The phenomena were very localized, and would almost inevitably have been overlooked except by very experienced lunar observers, but it did seem that there were too many reports to be dismissed out of hand. A few areas seemed particularly subject to the obscurations – notably Aristarchus (much the brightest crater on the Moon), Alphonsus in the Ptolemæus chain, and Gassendi on the edge of the Mare Humorum. Significantly, all these areas were rich in rills.

When I became Director of the Lunar Section of the British Astronomical Association, in the 1950s, I organized a network of observers to see whether we could obtain definite proof of the reality of TLP. One of our main areas was the floor of Alphonsus. I had been in touch with various professional astronomers, and in November 1958 one of them, the late Nikolai Kozyrev, made an observation of tremendous significance. He was using the 50-inch reflector at the Crimean Astrophysical Observatory when he recorded a temporary red patch inside Alphonsus, and confirmed it spectrographically. Kozyrev believed that the event was due to gaseous emissions from below the Moon's crust, and that the redness indicated high temperature. It now seems certain that an increase of heat was not the cause, but the observation itself was certainly valid, and at once TLP became officially 'respectable'. In 1963 some professional astronomers at the Lowell Observatory, in Arizona, reported red patches near Aristarchus, and in April 1966 redness was also seen by members of the British network, including myself.

Interest was aroused, and many observers joined in the hunt. At the Lunar and Planetary Laboratory at Tucson, Arizona, Barbara Middlehurst began collecting all past TLP reports, of which there were many; I had started a similar investigation, and when we compared notes we found that our results were almost identical, so that NASA published them in a joint report. At the meeting of the International Astronomical Union in 1969 we suggested that there could be a connection between our TLP sites and 'moonquakes' – slight disturbances in the Moon's crust; when the findings of the seismometers left on the surface by the Apollo astronauts came through, we were shown to be right.

TLP are not distributed at random. They are most often seen in a few definite areas, and also round the boundaries of the regular maria; some are red, others colourless, so that they are detectable only by a slight blurring of detail which would normally be expected to be visible – as in the floor of Plato, for example. They are also more common near the time of perigee, when the Moon is at its closest to the Earth and the crust is under maximum strain.

It is fair to say that some astronomers still doubt the reality of TLP, but it does seem that the evidence is very strong indeed, and the theory of gaseous emission is reasonable enough. What we really need, of course, is spectrographic confirmation; up to now, Kozyrev's observation of 1958 remains the only really reliable case, though a photograph of a bright outbreak near Proclus was obtained by the Greek astronomer G. Kolovos in 1985. The work continues, and at present (1989) one of the British Astronomical Association team, G. North, is taking regular lunar spectra with the 30-inch telescope at the Royal Greenwich Observatory, Herstmonceux.

Use can be made of a device known as a Moon-Blink. This takes the form of a rotating wheel, with red and blue colour filters, placed just on the object-glass side of the eyepiece (or the mirror side, with a reflector). A red patch on the Moon will show up as a dark feature when observed through the blue filter, but will be masked by the red filter. Rotating the wheel, the observer uses red and blue in quick succession, so that any red patch will show up as a 'blinking' spot. The method is surprisingly sensitive, and phenomena can be detected which would otherwise be missed – but, of course, it is confined to red events, and we have found that by no means all TLP are red.

It is not hard to make a Moon-Blink device, and it can be used with a modest telescope, though I would not be happy with any aperture smaller than 8 inches.

The trouble about TLP is that they are so elusive, and also uncommon – the observer may go for many months without seeing anything suspicious. Also, the greatest care is needed, and a bad observation is worse than useless, because it distorts the analyses. If you see what you think may be a TLP, check all other formations in the area – and in other areas – to make sure that you are not being deceived by any effects of the Earth's atmosphere. I do not regard any TLP report as valid unless it is confirmed, quite independently, by at least two observers at two different sites.

It is also worth investigating what are termed albedo variations. For example, the floor of Plato sometimes appears much darker than at other times, for reasons which are not fully understood. Generally, the darkest patch on the Moon is an area on the floor of the great walled plain Grimaldi, close to the western limb, but this is not always so. Variations are also found with some of the most brilliant features, such as Proclus just outside the Mare Crisium. Aristarchus is of special interest, and can often be seen clearly even when on the side of the Moon lit only by the earthshine.

No doubt the time will come when lunar bases will be set up, and many of the present-day problems will be cleared up. Yet even when this has happened, the Moon will retain its fascination for those of us who can never hope to travel in space. It is unique; it can never lose its appeal or its romance. There will always be endless pleasure to be gained from turning a telescope toward it and looking at the craters, the mountains and the valleys, learning how to recognize them and watching their shadows shift and change as the Sun rises over them.

7

Occultations and Eclipses

Occultations and eclipses have been described as the celestial equivalent of hide-and-seek. They are fascinating to watch, and they also provide the amateur with scope for some really useful work.

Because the Moon is so close to us on the cosmical scale, it moves against the starry background at a relatively high speed. Sometimes it must, of course, pass in front of a star and hide, or occult it. These occultations are common enough, but so far as bright stars are concerned they are not so frequent as might be thought. People tend to overestimate the apparent size of the Moon in the sky; artists will usually draw it as grotesquely large, whereas the lunar disk could actually be covered by a one-inch disk held at nine feet from the eye. (If you doubt me, try it for yourself!) Remember, too, that the Moon always keeps to the region of the Zodiac, and the only really brilliant stars which can ever be occulted are Antares, Aldebaran, Spica and Regulus.

A star appears to all intents and purposes as a point source of light. Therefore it shines steadily until the moment of occultation, when it snaps out like a candle-flame in the wind. One instant it is there; the next, it has gone. This is one proof that the Moon has no atmosphere, since a blanket of air round the limb would make the star flicker and fade for a few seconds before vanishing – as does in fact happen when a star is occulted by a planet such as Venus.

Seen in a telescope, an occultation is quite spectacular. The star seems to creep toward the Moon, though actually it is the Moon which is moving; the inexperienced observer is bound to feel that the star hangs close to the limb for a long time. Then the point of light will disappear, and the watcher who blinks his eyes at the wrong moment may easily miss the occultation completely. The emersion, at the opposite limb, is equally abrupt.

Before full moon, occultations take place at the unlit hemisphere, though very often there is enough Earthshine to tell the observer just where the limb is. After full, the star enters at the sunlit limb, and timings are less easy unless the star is a bright one.

Occultations are more important than might be thought. They can be predicted, but the Moon's apparent path in the sky is still not known with absolute precision. If an occultation is timed accurately, it gives the true position of the Moon's limb at that moment.

This is work that the amateur can do, but it needs the greatest care, and no observation is of much use unless it is correct to one-tenth of a second at least. The method I use is to start my stop-watch at the moment of occultation, and then check the watch against a time-signal as soon as possible. Radio signals are excellent, but the Speaking Clock is usually good enough. Unfortunately there can be no second chance, as I once found out to my cost. I had made what I thought was a good timing of an occultation in which I was particularly interested, but as I stepped away from my telescope I fell over the cat and re-started my watch, thereby putting paid to the entire operation. Clouds, too, always seem to roll up at exactly the wrong moment.

Photoelectric timings are naturally better than visual estimates, but this needs special equipment, and since I have never been a photoelectric observer I will say no more about it here.

When an occultation report is drawn up, remember to give the name or number of the star, the time of occultation to the nearest tenth of a second or better, the exact latitude and longitude of the observing station, its height above mean sea level of the observing station, and atmospheric conditions, plus type and magnification of the telescope used. A slight error in position can cause errors in the final computations. On one occasion I sent in some results, only to be told that there was something wrong with them. After some consideration, I found out what it was. Instead of using my 15-inch reflector, as I usually do, I had used my 12½-inch. The distance between the two telescopes is only about a couple of hundred yards, but in the reductions it showed up.

I remember, too, that when I was timing some occultations of stars in the Pleiades cluster I was surprised to find that one of the stars 'faded out', taking about 0.5 seconds to disappear instead of vanishing instantaneously. The star proved to be close double, so that the occultation took place in two stages. It is always worth watching for these fading occultations, though in most cases at least the double-star explanation is the right one.

Of particular value are 'grazing occultations', when the star skirts the northern or southern limb of the Moon. Because the lunar limb is mountainous, there may be several appearances and disappearances, and timings will give very useful information about the contour of the Moon's limb at the point of graze. Observations of this kind can be made only over a track little more than a thousand yards wide, and this is where amateurs can help, because they can take portable telescopes to the relevant area. Clearly it is best to work with a team; for a good 'graze' you really need several observers and several stop-watches. Predictions for grazes are given in many yearly astronomical handbooks and periodicals. (Fig. 7.1.).

Now and then unexpected occultations take place – unexpected in the sense that they are not listed in the tables supplied to the observer. Always be on the alert for anything unusual.

Occultations of planets are uncommon, but they do occur now and then. Because a planet appears as a disk, rather than a point source, the immersions and emersions take a few seconds, and it cannot really be said that much of value can be accomplished, but the phenomena are fascinating to watch. I have seen only one graze occultation of a planet – Venus, on 4 October 1980; with a team of observers I drove from Sussex up to a deserted airfield in Lincolnshire which marked the centre of the track. It will be many years before anything of the sort happens again.

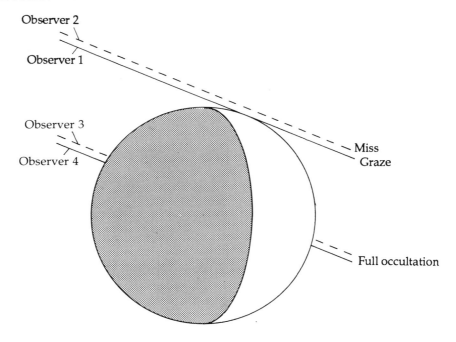

Fig. 7.1. Occultations. Observer 1 will see a 'graze' as the star skirts the lunar limb. Observers 3 and 4, in a different site, will see a full occultation.

Because the planets appear so much smaller than the Moon, they do not occult stars nearly so often, but now and then there is something exceptional to see. For instance, this happened on 7 July 1959, when Venus passed in front of the bright star Regulus. The occultation took place in broad daylight, but I was able to observe it with a 12-inch reflector, and saw the star flicker for almost a second before immersion – giving a clue to the height of the planet's atmosphere; remember that in 1959 we knew a great deal less about Venus than we do today. (The next occultation of Regulus by Venus will take place on 1 October 2044. I am afraid I will be unable to repeat my observation, unless of course I am lucky enough to live to the advanced age of a hundred and twenty-one!)

A really important occultation took place in 1988. The planet concerned was far-away Pluto, which looks like little more than a speck of light, and is attended by a satellite, Charon. Pluto was thought to have an atmosphere, but its extent was not known. When Pluto occulted a star on 9 June 1988, astronomers in New Zealand and Australia found that there was pronounced fading before and after the actual event, and from this it became clear that Pluto's atmosphere is much more extensive than had been previously thought. The point here is that amateurs took part with great success – notably Colonel Arthur Page from his private observatory at Mount Tamborine in Queensland. Of course, he has equipped himself with photoelectric equipment of full professional standard, but he is an amateur none the less – and he was by no means the only amateur taking part in this particular programme.

On very rare occasions, one planet may occult another. Thus Venus occulted Jupiter on 3 January 1818, and will do so again on 22 November 2065. There can sometimes be close conjunctions, and these are fascinating to watch and to photograph, though it cannot honestly be said that they are of any scientific value.

There is one type of observation which has come very much to the fore in recent years, thanks mainly to the work of Gordon Taylor at the Royal Greenwich Observatory. It is never easy to make an exact measurement of the apparent diameter of a very small object, such as an asteroid or one of the remote planets (Uranus, Neptune or Pluto); but if one of these objects occults a star, the duration of the phenomenon will allow the diameter of the occulting body to be worked out. Excellent results have been obtained with several asteroids, and, incidentally, it was an observation of this kind which led to the discovery of the ring-system of Uranus, about which I will have more to say in Chapter 11.

Of course, one has to wait for Nature to provide a suitable occultation, and Nature can be very unhelpful; also, the observer has to be in exactly the right place at exactly the right time. But the phenomena do occur; thus in 1987 there were no less than 25 predicted occultations of stars by asteroids of magnitude 13.5 or brighter.

If the asteroid is fainter than the star, and perhaps even below the limit of the observer's telescope, the star will simply disappear as the asteroid passes over it. If both star and asteroid are bright enough to be seen, there will be a change in the total magnitude when the occultation takes place – the important measurement being that of duration. A typical case was that of 8 December 1987, when the asteroid Bamberga, of magnitude 10.5, occulted a star of magnitude 9.8, and the event, visible from parts of Asia and the USA, lasted for a full half-minute.

Here, too, predictions are issued by the leading astronomical societies, and amateur collaboration is of immense value, because amateurs are mobile. Of course photoelectric measurements are more accurate than visual observations, but even the user of an ordinary telescope and a stop-watch can produce results of great value.

Next let me turn to eclipses, which are nothing if not spectacular. A solar eclipse is really an occultation of the Sun by the Moon, but a lunar eclipse is quite different; it is caused when the Moon passes into the cone of shadow cast by the Earth.

The Moon, as we know, has no light of its own, so that when it passes into the shadow it turns a dim, often coppery colour (Fig. 7.2). The main cone, shaded in the diagram, is known as the umbra, while to either side of it is the penumbra, caused by the fact that the Sun is a disk rather than a point source. (Note that in this context, umbra and penumbra have nothing to do with sunspots.) The diagram is not to scale, but it will, I hope, show what happens. The average length of the Earth's shadow is 850 000 miles, which is more than three times the distance of the Moon from the Earth, so that at the mean distance of the Moon the cone has a diameter of about 5700 miles. Obviously, the umbra is extensive enough to cover the Moon completely, and totality may last for as long as a hour and three-quarters, remembering that in the course of an hour the Moon moves across the sky by an amount which is slightly greater than its own diameter.

Every scrap of direct sunlight is cut off from the Moon as soon as it passes into the umbra, but the Moon does not (usually) disappear completely, because a certain

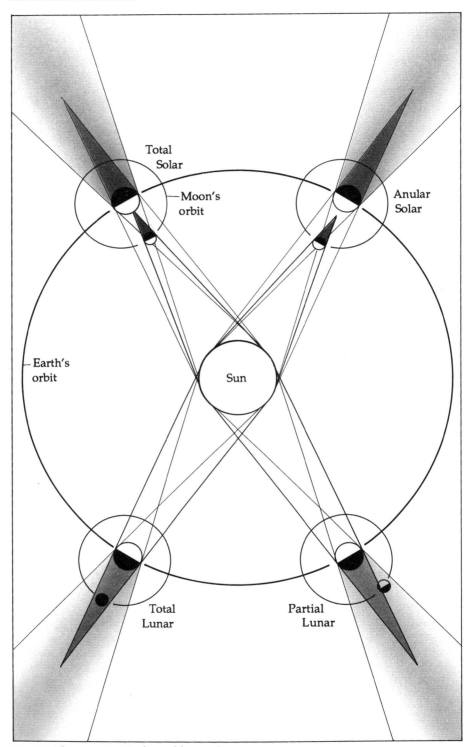

Fig. 7.2. The geometry of solar and lunar eclipses.

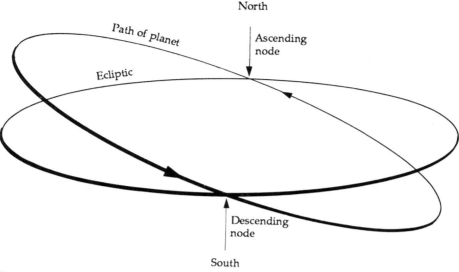

Fig. 7.3. Ascending and descending nodes.

amount of sunlight is bent on to its surface by way of the Earth's atmosphere. While inside the shadow the Moon often shows beautiful colour effects, and not all eclipses are equally dark – everything depends upon conditions in our own air at the time. It is reliably reported that during the eclipses of May 1761 and June 1816 the Moon could not be found even with a telescope, whereas in 1848 the totally eclipsed disk was so bright that many people refused to believe that anything unusual was happening.

The French astronomer André Danjon once produced an eclipse scale, ranging from 4 (bright) to 0 (very dark). I have seen a number of totalities, some much darker than others – as on 25 June 1964, when I could not see the Moon at all; this was because of volcanic dust which had been sent into the upper atmosphere by a violent eruption in the East Indies.

If the Moon does not enter wholly into the umbra, the eclipse is partial, while at other times it is merely penumbral. It is often said that a penumbral eclipse produces an effect too slight to be seen with the naked eye, but I do not agree; generally speaking, the dimming is obvious enough.

Two more things are clear from the diagram. First, a lunar eclipse must be visible over one complete hemisphere of the Earth, provided that clouds do not hide it; if it is total anywhere, it must be total everywhere. Secondly, an eclipse can happen only at full moon.

Lunar eclipses are so obvious that they must have been noticed from very early times, and records of them go back to several thousands of years BC. Were the Moon's orbit not tilted across the ecliptic, there would be a total eclipse every month, but the inclination of the Moon's path is enough to prevent this from happening. Imagine two hoops hinged along a diameter, and crossing each other (Fig. 7.3). The points at which they cross are called the 'nodes', and unless full moon occurs very near a node it will miss the shadow altogether.

It so happens that the Sun, Moon and nodes return to almost the same relative positions after a period of 18 years $10\frac{1}{4}$ days, so that any particular eclipse will be followed by another one 18 years $10\frac{1}{4}$ days later. This is the so-called Saros period, known to the astronomers of Ancient Greece and used by them to make eclipse predictions. The Saros is not exact, but it is good enough to be useful.

It is interesting to note the various colours seen during the progress of the eclipse, and to time the moments when the shadow reaches or leaves definite features such as craters. The effects on the Moon itself must be very marked. The lunar surface is poor at retaining heat, so that the temperature there drops sharply as soon as the shadow is cast; the fall may amount to 100°C in the course of an hour. Some areas cool down less rapidly than their surroundings, and are known, rather misleadingly, as 'hot spots'; ray-craters such as Tycho come into this category. Temperature measurements are still of considerable importance, but I do not propose to discuss them here, because the equipment needed is beyond the range of the average amateur.

Whether any visible effects can be tracked down in surface features is much more dubious. One area studied a few decades ago was the famous (or notorious!) Linné, in the Mare Serenitatis. It was suggested that the white patch surrounding Linné became larger during the eclipse, presumably because of the sudden drop in temperature. Frankly, I am highly sceptical; I have made measurements at various eclipses with completely negative results, and neither have I found any effects upon other 'suspect' areas, such as the brilliant Aristarchus and the interiors of dark-floored walled plains such as Plato and Grimaldi. Variations are always worth looking for during eclipses, but in my view there is little chance of finding anything definite.

With eclipses of the Sun, we are back to the occultation principle, since a solar eclipse is caused simply by the Moon passing between the Sun and the Earth.

The Moon's diameter is only about $\frac{1}{400}$ that of the Sun, but it is also 400 times closer, so that in our skies the two bodies look almost exactly the same size. When the Earth, Moon and Sun line up, with the Moon in the mid-position, the shadow cast by the Moon is just long enough to touch the Earth's surface, and at totality the brilliant solar disk may be blotted out for as long as $7\frac{1}{2}$ minutes, though admittedly most total eclipses are much briefer. The zone of totality can never be more than 169 miles wide, so that once again the observer has to be in the right place at the right time. From any particular location, solar eclipses are much less common than eclipses of the Moon, which, as have seen, are visible over a complete hemisphere. From England, the last total solar eclipse was that of 1927; the next will be on 11 August 1999, when the central track will pass across Cornwall.

A solar eclipse can happen only at new moon, and then only if new moon occurs near a node; otherwise, the Moon will pass unseen either above or below the Sun in the sky. This is not always appreciated. For example, in the first edition of the classic novel *King Solomon's Mines*, H. Rider Haggard described a full moon, a solar eclipse, and another full moon on successive nights. When someone pointed out that he had made a slight mistake, he corrected the second edition, prudently turning his solar eclipse into a lunar one!

The Saros period is valid, but the rough-and-ready method of forecasting is less

Fig. 7.4. Annular eclipse; my photograph, from Thera (Mediterranean) in 1975. A ring of the Sun was left showing round the dark disk of the Moon.

satisfactory. For instance, the 'return' of the 1927 eclipse took place in 1945, but in this year the zone of totality lay further north, and missed England altogether, though a partial eclipse could be seen.

To either side of the track of totality the eclipse will be partial, while some eclipses are not total anywhere on the Earth. There is also a third kind of eclipse – the annular (Latin *annulus*, a ring). When the Moon is near apogee, it appears smaller than the Sun in the sky, and cannot cover the whole of the disk. When the three bodies line up under these conditions, a bright ring of the Sun is left showing around the dark mass of the Moon. Sometimes an eclipse may be total over part of the central track and annular over the rest (Fig. 7.4).

Partial and annular eclipses are interesting to watch, but, as always, remember to take care. Even when most of the Sun is hidden, it is still most unsafe to look direct either with binoculars or with a telescope. The slightest sliver of direct sunlight remaining is enough to damage the observer's eye in a matter of seconds, and at every solar eclipse hospitals report cases of people who have injured themselves by taking 'just one quick peep'. If you want to watch the eclipse with the naked eye, use a very dark filter, and do not stare for more than a few moments at a time.

A total eclipse is unquestionably the finest display in all Nature (Fig. 7.5). As the

Fig. 7.5. A total eclipse of the Sun, showing the corona. I took this photograph from Java, on 11 June 1983, using a 400 mm telephoto lens.

Moon's shadow sweeps on, the light fades, and at the instant when the last of the photosphere is covered the atmosphere of the Sun leaps into view. There are the magnificent red prominences; there is the chromosphere, and there is the 'pearly crown' or *corona*, sometimes fairly regular in outline and sometimes sending out streamers across the sky. Totality is all too brief; then there is a flash as the Sun starts to reappear, producing the wonderful 'diamond ring' effect, and then, with bewildering suddenness, the light comes back and the corona and prominences fade away.

The prominences are visible with the naked eye only during totality, though with special instruments based upon the principle of the spectroscope they can be observed at any time. They are made up of incandescent gas, and are of enormous size; the length of an average prominence may be well over 120000 miles. Many are associated with spot-groups, and they too tend to follow the eleven-year cycle.

Quiescent prominences are relatively calm, as their name suggests, and may last for several weeks before either breaking up gradually or else being violently disrupted. Active prominences may be likened to tall tree-like structures, from the tops of which glowing streamers flow out horizontally and then curve downward toward the photosphere. They move rapidly; some of the blown-off material has been known to move at over 400 miles per second.

The corona forms the Sun's outer atmosphere. It is much more extended than the chromosphere, and is composed of tenuous gas, millions of times less dense than the air that you and I are breathing. It stretches outward for many millions of miles, although because of its low density and indefinite boundary it is not possible to give an exact figure for its 'depth'.

The first total eclipse which I saw was that of 30 June 1954. It was just total off the coast of North Scotland, so that a large partial eclipse was seen in England, and caused general interest even among people not usually astronomically-minded. The main track crossed Scandinavia, where many astronomers gathered. The combined Royal Astronomical Society and British Astronomical Association expedition, of which I was a member, made its headquarters at the little Swedish town of Lysekil, along the coast from Göteborg. Our arrival there coincided with the Midsummer Festival. It also coincided with a burst of torrential rain.

On 30 June, most observers collected their equipment and drove to Strömstad, at the exact centre of the track, almost on the Norwegian frontier. The site selected was a hill overlooking Strömstad itself, and by noon it was littered with equipment of all kinds: telescopes, spectroscopes, thermometers, cameras, and even a large roll of white paper that I had spread out in the hope of recording shadow bands. These shadow bands are curious wavy lines which appear just before totality. They are due to atmospheric effects, but are hard to photograph, and the opportunity seemed too good to be missed. I remember that one eminent astronomer went to the trouble of photographing my apparatus, commenting dryly that although he might live to see another total eclipse, he would never again see so peculiar an arrangement.

The early stages of the eclipse were well seen. Five minutes before totality, everything became strangely still, and over the hills we could see the approaching area of gloom. Then, suddenly, totality was upon us. The corona flashed into view round the dark disk of the Moon, a glorious aureole of light which made me realize the inadequacy of a mere photograph. The sky was fairly clear; although a thin layer of upper cloud persisted, it did not cause any real trouble.

It was not really dark. Considerable light remained, and of the stars and planets only Venus shone forth. Yet the eclipsed Sun was a superb sight, with brilliant inner corona and conspicuous prominences. The $2\frac{1}{2}$ minutes of totality seemed to race by. Then a magnificent red–gold flash heralded the reappearance of the chromosphere; there was the momentary effect of the 'diamond ring', and then totality was over, with the corona and prominences lost in the glare and the world waking up once more. I have seen four totalities since then, but I will never forget the thrill of that first experience, even though it happened more than 30 years ago.

It is also true that the eclipse observer must be ready for any eventuality. I found this out during the eclipse of 18 March 1988, which I observed from the Philippines, or more precisely, from Talikud Island, in Davao Bay.* The weather was expected to be good, but on the morning of 18 March the sky was overcast, and as we set up our telescopes and cameras we were in a decidedly gloomy mood. Ten minutes before totality there was even a shower of rain, and most of us abandoned all hope.

---

* Initially we were threatened with some non-astronomical problems. In the Philippines there are several groups of terrorists, and as they tend to blow things up some members of our party were somewhat apprehensive. However, they very sportingly issued a combined Press statement that they were not going to interfere with the visiting astronomers – and they didn't!

Fig. 7.6. Total eclipse, 1988 which I photographed from Talikud Island. The corona was very striking.

Then, 30 seconds before totality, the clouds thinned, and the crescent Sun could be seen above the palm-trees; it was a breathtaking spectacle. There was no time to make adjustments to allow for the residual cloud, and I simply 'clicked off' a series of exposures, hoping that one of them would turn out to be right. I did, in fact, obtain one really good picture. Had I paused to try for perfect focus and perfectly correct exposure, I am sure I would have failed – as, alas, some members of our party actually did.

Because total eclipses are so spectacular, and because they give scope for work which cannot be carried out at any other time, astronomers are always ready to go on long journeys to observe them – often, to be thwarted by clouds at the critical moment. In general I have been busy with television and broadcast commentaries during my eclipse trips, but I have also tried to photograph the surrounding sky to search for unexpected comets. The classic case of this goes back to 1882 (rather before my time!). Photographs were taken of the total eclipse of that year, and, to the general surprise, they also showed a bright comet close to the Sun. It had never been seen before, and it was never seen again, so that this is our only record of it. I have always hoped to make a discovery of the same kind, though so far I have had no luck.

There are many special investigations to be undertaken during a total eclipse, but most of them require elaborate equipment, and most amateurs will content themselves with taking photographs. I do not pretend to be a serious photographer, but I have taken some eclipse pictures which show the corona well (Fig. 7.5), and I have given some details of exposures and technique in Appendix 29. Meanwhile, I suggest that if you ever have the chance to watch a total eclipse, you should take it. We must be thankful that by sheer coincidence, the Sun and Moon look almost exactly the same size in our sky. Were the Moon a little smaller, or a little further away, we could never be treated to the naked-eye spectacle of the prominences and the corona.

8

Glows in the Sky

On the evening of 26 July 1938 I went into the garden of my Sussex home to be greeted by a remarkable spectacle. The whole of the western sky was lit with a vivid red glow, with green and silver patches appearing here and there. For a moment I wondered whether the whole of London was on fire, but then I realized what I was seeing. It was no fire; it was a brilliant display of aurora borealis.

Auroræ or polar lights, have been known from very early times, and are so common in high latitudes that a night in north Norway or Antarctica would seem drab without them (Fig. 8.1). They occur at heights ranging from over 600 miles down to below 60; sometimes they take the form of regular patterns, while at others they shift and change rapidly, showing brilliant colours and providing a spectacle which is second in glory only to a total solar eclipse.

Since meteorology is the science of the atmosphere, it might be thought that auroræ are outside the scope of the astronomer. Yet their cause is to be found not on Earth, but in the Sun. Active areas on the Sun, often connected with flares, send out electrified particles, and it is these particles which enter our air and produce the glows. Predictably, auroræ are most common near the time of sunspot maximum, and a major flare occurring near the centre of the disk is often followed a day later by a bright aurora. Since the particles must therefore cover the 93 million-mile gap in about 24 hours, it follows that their speed must be of the order of 1000 miles per second.

The fact that auroræ occur mainly in high latitudes is because the electrified particles are attracted toward the Earth's magnetic poles. They are commonest around geomagnetic latitude 68 degrees north or south – aurora borealis in the northern hemisphere, aurora australis in the southern. On average, auroræ of some kind or other are seen on 240 nights per year in Northern Alaska, North Norway, Canada and Iceland; 25 nights per year along the Canada–United States border and in Central Scotland, and one night per year in Central France. On occasion they may extend further toward the equator. It is on record that an aurora was once seen from Singapore (latitude 1°20′ N). Data from the southern hemisphere are much less complete, but auroræ have been recorded in Australia and South Africa, and they are not too uncommon in the South Island of New Zealand.

Scientifically, auroræ are important not only because of their link with the Sun, but

Fig. 8.1. Aurora Borealis, seen from Hampshire, England. (Photograph by Ron Arbour.)

also because they provide information about conditions in the upper atmosphere. It is therefore useful to observe them whenever possible, and to make estimates of their positions against the starry background, so that their heights may be worked out. Amateurs can play a major rôle here; for example, a full-scale survey is organized by the Aurora Section of the British Astronomical Association, which has members all over the world. Observers taking part are asked to fill in forms giving details of the presence or absence of aurora, coupled with notes of any displays that may be seen.

One never knows quite what an aurora is going to do next, but a great display often begins as a glow on the horizon, rising slowly to become an arc. After a while, the bottom of the arc brightens, sending out streamers, after which the arc itself loses its regular shape and develops folds like those of a radiant curtain. If the streamers extend beyond the zenith, or overhead point of the sky, they converge in a patch to form a corona (not, of course, to be confused with the solar corona). Flaming auroræ are composed of quickly-moving sheets of light, while 'ghost arcs' may persist long after the main display has ended. The whole phenomenon may extend over many hours.

For observing auroræ, by far the best instrument is the naked eye, coupled with a red torch and a reliable watch. Binoculars are of little help, and telescopes absolutely useless. Points to note are the bearing of the centre of the display, reckoned in degrees (0–360) from north round by east; the various forms seen, such as arcs, curtains, draperies and flaming surges, colours, and duration. Times should be taken to the nearest minute. There is obvious scope for the photographer, and spectroscopic work is valuable, but simple naked-eye observation is not to be despised.

Extremely low auroræ have been reported now and then, but with no certainty.

There have also been reports of sounds accompanying auroral displays — either crackling or hissing. I tend to be sceptical about this, though I admit that the reports are so numerous that they cannot be rejected out of hand. Odours have also been reported, though smelly auroræ seem even less plausible than noisy auroræ!

The most brilliant aurora of recent years was that of 13 March 1989, associated with the huge sunspot group which I mentioned earlier. As seen from Britain it was superb; from my Sussex home I missed the best of it, but from some parts of England it cast shadows, and from Stoke-on-Trent Paul Doherty, who took some superb pictures of it, reported that it was so brilliant that one could read a paper by its light. It was seen as far south as France. At the same time, there was an equally good display of aurora australis. We cannot tell when another comparable aurora will be seen, but it may happen at any time.

I must also make brief mention of noctilucent clouds (Fig. 8.2), which are quite unlike ordinary clouds. They form at altitudes of 50 miles or so, above 99 per cent of the atmosphere, where the pressure is only a millionth of that at sea level. Again they are mainly phenomena of high latitudes, mainly in zones between 60 and 80° N or S, so that they cannot often be observed from parts of Britain. They are visible only when the Sun is between 6 and 12 degrees below the horizon. Any higher, and the sky will be too bright; any lower, and they will be covered by the Earth's shadow. In general they are of a delicate silvery-blue colour, often with a 'herring-bone' structure produced by delicately interwoven bands.

The cause of noctilucent clouds is not known with any certainty. It may be that they condense around specks of meteoritic dust; it may be that they condense round particles sent up by volcanic eruptions — but it is probably significant that they are commonest at times when auroræ are least evident, around the time of solar minimum. They are quite easy to photograph, and are always worth looking for.

Another 'sky glow' is the Zodiacal Light, seen as a faint luminous cone rising from the horizon after sunset or before sunrise. It can be quite prominent from countries where the air and dust-free, though from Britain it is usually hard to see. It extends away from the Sun, and is generally visible for only a short time during late evening or early dawn. The Zodiacal Band, a faint, parallel-sided extension of the cone, may extend right across the sky to the far horizon, though it is so dim that it seldom to be observed at all except from the tropics.

The Zodiacal Light originates well beyond the top of the atmosphere. It is due to particles scattered along and near the main plane of the Solar System; the diameters of these particles are of the order of 0.1–0.2 of a micron. (One micron is equal to one-millionth of a metre.) The best times for observing the Light are evenings in March and mornings in September, because at these times the ecliptic is most nearly perpendicular to the horizon, and the cone is higher in the sky.

Since the Zodiacal Light is faint, its intensity is not easy to estimate. The best method is to compare it with a definite region of the Milky Way. The width of the base, in degrees, should also be noted. Though the Light is predominately white, a pinkish or at least warmish glow has been reported in its lower parts, and should be looked for.

Fig. 8.2. Noctilucent clouds, seen from Dundee, Scotland. (Photographed by K. Kennedy.)

Last and most elusive of these glows is the Gegenschein, a faint patch of radiance always exactly opposite to the Sun in the sky. To see it at all, you need near-perfect conditions, with a black sky and a complete absence of light pollution. From Britain I have seen it only once – in March 1942, when the whole country was blacked out as a precaution against German air-raids.

The best chances of observing it are when the anti-Sun position is well away from the Milky Way, i.e. in February–April and September–November, and the anti-Sun position is high, at local midnight. It may then appear oval in shape, with a diameter roughly forty times that of the full moon. It too is caused by sunlight reflected from tiny particles in the main plane of the Solar System, and on occasion a faint extension known as the Zodiacal Band may link it to the Zodiacal Light.

For all these observations, one point should be borne in mind: Never begin work before you have made your eyes thoroughly accustomed to the dark. To come outdoors from a brilliantly-lit room and expect to see an auroral glow or the Zodiacal Light straight away is fruitless, and it is usually wise to walk around for the best part of an hour before starting your programme, though the exact period is bound to vary with different people. For recording observations, use a torch with a red bulb.

Here again, the amateur has a major part to play. There is no need to wait years for a great aurora; studying the fainter lights and glows is fascinating, and it is a pity that town-dwellers never have a chance to see the ghostly beauty of the Zodiacal Light.

9

The Nearer Planets

The planets are of special interest to the owner of a small or fair-sized telescope (by which I mean an aperture of anything between 3 and 18 inches). Most of them show definite surface features; they are always changing, and there are still opportunities for doing some useful research.

The first of them, reckoning outward from the Sun, is Mercury, undoubtedly the least promising of the naked-eye planets. It has a diameter of only 3030 miles, so that it is not a great deal larger than the Moon, and it never comes much within fifty million miles of us. Because it always stays close to the Sun in the sky, it is not easy to see with the naked eye except when it is at its very best. Using a telescope or binoculars to sweep for it when the Sun is above the horizon is emphatically not to be recommended. To locate Mercury in the daytime, you need an equatorial, driven telescope with accurate setting circles. Otherwise, you will have to make do with a planet which is low over the horizon, so that conditions will be unsteady. A good plan is to find it in the dawn sky before sunrise, and then keep it in the telescope field until it has reached a reasonable altitude. Unlike Venus, Mercury is at its brightest when at the gibbous phase. It is much less reflective than Venus, mainly because its weak escape velocity means that it has no appreciable atmosphere.

Before the Space Age, the best map of Mercury was that drawn by the Greek astronomer Eugenios Antoniadi, who had the use of the great 33-inch refractor at the Meudon Observatory in France (a telescope which I know well, as I carried out extensive lunar mapping with it in the pre-Apollo period). Antoniadi, probably the best planetary observer of the time, showed shadings and brighter regions, and believed that the rotation period was 'captured', so that the Mercurian day and year were equal at 88 Earth-days. This would mean that the same face would be turned to the Sun all the time; there would be an area of permanent sunlight, an area of permanent night, and only a narrow 'twilight zone' between over which the Sun would rise and set. In this Antoniadi was wrong. The real rotation period is only 58.6 days, so that every part of Mercury is in sunlight at one time or another – though it so happens that whenever the planet is best placed for observation, the same side is turned toward the Earth (a state of affairs which may or may not be due to coincidence). Then, in 1973–74, came the flight of the unmanned probe Mariner 10, which made three active passes of Mercury before its transmitters failed, and sent back detailed pictures showing mountains, craters, plains and valleys. In many ways the Mercurian surface is very like that of the Moon, though there are marked differences in detail.

Obviously, these features are not visible with telescopes of the size used by amateurs, and even Antoniadi's map was so inaccurate that his nomenclature had to be abandoned. Whether any surface features at all can be seen with, say, a 15-inch reflector is debatable. On occasions I have suspected some, but with no confidence. All in all, the amateur must content himself with locating Mercury and following the changes in phase. This may not be of any scientific value, but I for one find it distinctly satisfying!

Sometimes Mercury can pass in transit across the face of the Sun. When this happens, it can be seen as a well-defined black disk, much darker than a sunspot. Transits are interesting, and projection with a small telescope such as a 3-inch refractor will show them well. Two more are due before the end of the century: on 6 November 1993, and 15 November 1999.

Mercury is so small and remote that even if it did not remain obstinately close to the Sun, we could hardly hope to find out much about it by observation from Earth. Venus, next in order, is as different as it could possibly be. It is almost as large as the Earth, with an escape velocity of over 6 miles per second. In size and mass, it may even be said that Venus and the Earth are virtual twins.

Venus is a splendid object with the naked eye, and at times it can even cast a shadow, but telescopically we have to admit that it is a disappointment. When at its most brilliant it shows up as a crescent; by the time of dichotomy (half-phase) it has already drawn away from us, so that its apparent diameter is much less, and when full it is on the far side of the Sun (Fig. 9.1). Altogether, Venus is a most infuriating object. Moreover, the disk appears virtually blank even with large telescopes. Vague, dusky shadings can be glimpsed often enough, but they are so diffuse that they are hard to define. We are looking not at the true surface of the planet, but at the upper layers of a dense atmosphere. The 'weather' on Venus is always cloudy.

Up to the end of 1962 our information about Venus as a world was remarkably slight. Large amounts of carbon dioxide had been detected spectroscopically in the upper atmosphere, but nothing definite was known about the lower layers, and even the length of the axial rotation period remained unknown. Visual observers did their best to follow the drifting shadings as they were carried slowly across the disk, but with scant success. Estimates of the rotation period ranged between 16 hours and as much as 224.7 Earth-days — which is also the length of Venus' orbital period, so that in this case the rotation would be 'captured', with the same face turned permanently sunward. Opinions as to the nature of the surface fluctuated wildly between a swampy, tropical hothouse, a planet completely covered with water, and an arid, fiercely-hot dust-desert without a scrap of moisture anywhere.

The only way to find out was to send space-probes. The first successful mission was that of 1962, when America's Mariner 2 by-passed the planet at 21000 miles and showed that the surface really is intolerably hot. Since then there have been various other probes, both American and Russian. Some have been put into closed paths

Fig. 9.1. Phases of Venus. At new, the dark side is turned toward us, though sometimes the planet appears as a ring because of the illumination of its atmosphere by the Sun. As the phase increases, the apparent diameter shrinks; when full (on the far side of the Sun) the apparent diameter is least.

round the planet, and have charted most of the surface by radar, while others have made controlled landings, sending back photographs of the surface and managing to transmit for an hour or two before being put out of action by the hostile conditions. In 1985 the Russians even used their Vega probes, en route to Halley's Comet, to drop balloons into Venus' atmosphere; the balloons floated around at various levels, sending back useful information.

Our modern view of Venus is not encouraging. The atmosphere is composed chiefly of carbon dioxide, and since this acts as a 'greenhouse' the surface temperature is over 900°F. The atmospheric pressure at ground level is more than 90 times that of our own air, and the attractive-looking clouds of Venus turn out to be rich in sulphuric acid, while the lowest layers of the atmosphere could well be described as corrosive, superheated smog, with winds which have tremendous force even though they are relatively sluggish. The surface is so hot that it glows, and the sky is orange, with the

Sun and stars permanently hidden. Venus is a world of rolling plains, highland areas and craters; there are deep valleys, and there are volcanoes which are almost certainly active. Frankly, the overall scene is very much like the conventional picture of hell.

The axial rotation period has proved to be just over 243 days. This is longer than the orbital period of 224.7 days, leading to a very curious sort of calendar. To confuse matters still further, Venus rotates in a retrograde direction – that is to say, from east to west, not west to east as in the case of the Earth or Mars. If anyone standing on Venus could see the Sun through those choking clouds, it would be found that the Sun would rise in the west, crawl across the sky, and set in the east almost 117 Earth-days later. Just why Venus behaves in this peculiar manner is not known.

Interestingly, the upper clouds have a rotation period of only four days, though they still move in a retrograde sense. This was established in the 1950s by French astronomers working at the Pic du Midi Observatory in the Pyrenees. The result seemed so unlikely that many people were sceptical about it; I admit that I was extremely dubious, but the French work was correct, and showed that even today the visual observer has a rôle to play. It was the French, too, who persistently saw the famous 'V-formation' of shadings which had been previously recorded by many observers.

Life on Venus? So far as we are concerned, we must, I feel, rule it out. Neither is there any hope of manned expeditions there in the foreseeable future. The Planet of Beauty is not nearly so welcoming as she looks.

There is not much point in observing Venus when it is low down and dazzlingly bright. Luckily, the planet is relatively easy to find in daylight. It becomes visible with the naked eye well before sunset during evening apparitions, and remains on view during morning elongations well after the Sun has risen. Moreover, it can be found telescopically at any time when it is not too far from the Sun. If you have a telescope with an equatorial mount, the easiest procedure is to aim the telescope at the Sun without looking through it, set the declination difference between the Sun and Venus, and then sweep away from the Sun by the difference in right ascension. Venus will probably be found with no trouble at all. Remember, though, always to sweep *away* from the Sun, and do not range aimlessly around.

Obviously it is best to find the planet with a low magnification. Then change to a higher power – as high as can still provide a crisp, clear image – and draw in the phase. Next, look for any shadings that may be visible, and draw them, though you will probably have to exaggerate them. Then, look for any bright areas; often there are polar caps – due, I hasten to add, to cloud structures and not to snow or ice! Very often, contrast will lead to the impression of a dark collar around the edge of the cap.

Filters can be a help. A yellow filter often enhances the shadings, and I also use red and blue filters, which produce alterations in the form and position of any features which may be visible. This presumably indicates real differences in structure and in the heights of the clouds.

Generally, a scale of two inches to the planet's full diameter is convenient, and the phase can be measured directly from the drawing. Note also the shape of the terminator or boundary between the sunlit and dark hemispheres of the planet; it may be a smooth curve, there may be irregularities (certainly not due to high mountains, as was once suggested), and near dichotomy the outline may be 'ogee'. Neither are the cusps always equally sharp. Very often you will find that one is much blunter than the other.

We know the orbit of Venus very accurately, and therefore we should be able to give the phase at any moment, but in fact theory and observation do not agree. When Venus is an evening object, and waning, dichotomy is always early; when it is a waxing object in the morning sky, dichotomy is late. The discrepancy may amount to several days. The phenomenon was first noted in the 1790s by Johann Schröter, and some years ago I referred to it as 'the Schröter effect', a term which has now become accepted by everyone. It is not confined to the time of dichotomy; it persists for much of the phase cycle.

Dichotomy provides the best opportunities for checking on the Schröter effect, but is is not an easy matter, partly because Spode's Law ensures that clouds will interrupt the series of observations at the critical moment, and partly because the terminator may appear sensibly straight for several days in succession. There is no chance that Venus is out of position, and therefore the discrepancy must be due to the planet's atmosphere, but we are not yet certain whether there is any regularity about it. This is research which the amateur can undertake with any telescope of over 3 inches for a refractor or over 6 inches for a Newtonian reflector. Incidentally it seems that Mercury, which has no atmosphere, does not show the Schröter effect.

Next, there is the vexed question of the Ashen Light. When Venus is a crescent, the unlit hemisphere can often be seen shining faintly. The same thing is seen with the Moon, but can easily be explained as being due to Earthshine; Venus, however, has no moon, and from a distance of well over twenty million miles the Earth certainly cannot produce the glow.

Various weird ideas have been proposed. (In the last century, a German astronomer who rejoiced in the name of Franz von Paula Gruithuisen maintained that it was due to festival illuminations lit by the local inhabitants to celebrate the election of a new Government.) But on the whole it seems that the most likely cause is electrical phenomena in Venus' upper atmosphere, roughly analogous to our auroræ, though it must be added that Venus has no detectable magnetic field.

Observing the Ashen Light is always difficult. Obviously it can be detected only when Venus is seen against a relatively dark sky, which means that the planet is bound to be low down. It is easy to be misled by contrast effects, and if the observation is to be reliable the bright crescent must be screened by a suitable occulting device.

A curved 'bar' is not satisfactory (try keeping the bright crescent directly behind it, and you will see what I mean). I therefore use an eyepiece which has part of the field blacked out leaving the rest clear; the boundary of the 'bar' has three curved sections of different sizes. The crescent is then placed just behind the section which is of the right size at that particular moment. If the unlit portion of the disk remains visible, then you may be sure that you really are seeing the Ashen Light.

Incidentally, the illusion that the unlit portion of the disk appears darker than the surrounding sky, when seen against a brightish background, is certainly due to contrast and nothing else.

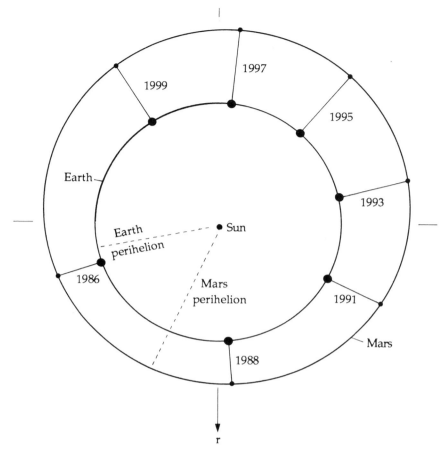

Fig. 9.2. Oppositions of Mars, 1986–99. During this period the most favourable opposition was that of 1988, when Mars was close to perihelion.

If electrical phenomena are responsible, we would expect that – like our aurora – the Light would be at its most prominent when the Sun is active. Unfortunately, the observations so far are not complete enough to provide proof. Observations cannot be made against a bright background, and when Venus is low down the conditions are bound to be poor. Also, there is no point in looking for the Light when the phase is greater than about 25 per cent, so that opportunities are limited. Yet it is a problem which is well worth following up.

All in all, Venus is more interesting to the telescope observer than might be thought at first sight – so do not neglect it, even though you may observe for many consecutive days before seeing anything definite enough to record.

Transits of Venus are irritatingly rare. They occur in pairs, separated by eight years, after which no more occur for over a century. In transit, Venus is said to be easy to see with the naked eye, but the next opportunity will not be until 2004. So let us turn now to Mars, which is a much more rewarding object inasmuch as it shows permanent surface markings.

Mars can approach us to within a distance of 35 million miles. Even so, it is always at least 150 times as far away as the Moon, and since it is a small world, with a diameter of only 4200 miles, its apparent diameter can never be more than about 24 seconds of arc. On the credit side, it is much more conveniently placed than either Mercury or Venus, and when near opposition it may be above the horizon all through the hours of darkness (Fig. 9.2). Obviously it can never appear as a half or a crescent; at its most gibbous the disk is still 85 per cent illuminated, so that it resembles the shape of the Moon a day or two from full.

The main trouble about Mars is that it comes to opposition only in alternate years; the mean synodic period is 780 days. Delicate details can be seen only for a few weeks to either side of opposition, unless a very powerful telescope is used. Neither are all oppositions equally favourable, because the Martian orbit is more eccentric than ours; the distance from the Sun ranges between 128.5 million miles and 154.5 million miles. Minimum distance from Earth occurs when Mars comes to opposition and perihelion at the same time; at aphelic oppositions, such as those of 1980 and 1995, the distance may never be reduced to as much as 60 million miles. The last really good opposition was that of September 1988 — particularly favourable for observers in Britain and much of the United States of America, because Mars was in the constellation of Pisces, well north of the celestial equator. Not until 2003 will there be another opposition as good as this.

When at its brightest, Mars is actually more brilliant than any other planet apart from Venus; it can surpass even Jupiter. When it is a long way away, the magnitude drops to around 1.8, and it is then only too easy to confuse Mars with a red star.

Mars has a revolution period of 687 Earth-days. This is equivalent to 668 Martian days or *sols*, because Mars spins more slowly than we do; the period is 24 hours 37.4 minutes. Because the atmosphere is thin, and the distance from the Sun greater than that of the Earth, it is logical to assume that Mars is cool, but it is not a permanently frozen world. At noon on the equator, at midsummer, the temperature may rise to at least 50°F. The nights, of course, are very cold indeed, so that from our point of view the Martian climate is decidedly uncomfortable.

When you first look at Mars through a telescope, you may well feel a sensation of anti-climax. Instead of a globe streaked with canals and blotched with patches of blue-green vegetation, you may be able to make out nothing except a tiny red disk crowned in the north or south by a white polar cap. It is only when you become really used to planetary observation that you will be able to make out the fine details. The markings on Mars are much less evident than the belts of Jupiter, the rings of Saturn or the phases of Venus; even the polar caps are beyond the range of a small telescope except when they are at their greatest extent. But the longer you look, the more you will see, and with practice you will soon learn how to recognize the various features, noting the slow drift as they are carried across the disk by virtue of the planet's rotation (Fig. 9.3). The shifts become quite noticeable over a period of only an hour or two.

The first good map of Mars was drawn in 1877 by the Italian astronomer G.V. Schiaparelli, using an 8¾-inch refractor. Schiaparelli charted the bright and dark areas very accurately, and gave them names which are still in use, though they have been

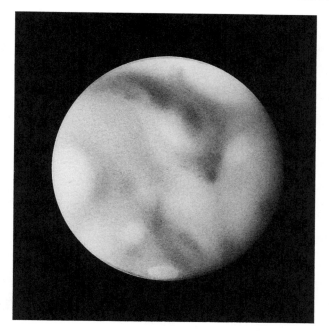

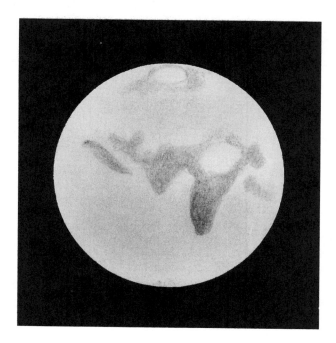

Fig. 9.3. (a) Mars; (Paul Doherty, 12 April 1982, 2150 GMT, 419 mm reflector, × 248.) (b) Mars: (Patrick Moore, 1988 September 19, 2235 GMT, 15-inch reflector, × 400.) Central meridian 278 degrees. Diameter 23″.5. Seeing II (Antoniadi scale). Syrtis Major is well seen.

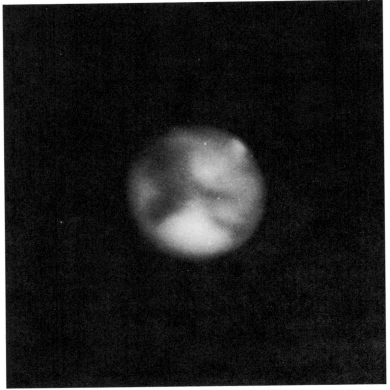

Fig. 9.4. Mars. Note the V-shaped Syrtis Major, and the south polar cap. (Photograph by Ron Arbour, 18 September 1988 with his 40 cm reflector, exposure ⅕ second.)

modified in recent years. Schiaparelli also drew a network of straight, artificial-looking lines which he called *canali* (channels), but which are always known as the Martian canals. Inevitably, the suggestion was made that these canals might be artificial. Percival Lowell, who built a major observatory at Flagstaff, Arizona mainly to study Mars, was convinced that the Red Planet supported an advanced technical civilization, and that the canals made up an irrigation system designed to carry water from the polar snows through to the arid deserts of the equator. One enthusiastic Lowell supporter even went so far as to work out the power of the pumps needed!

Lowell's views met with considerable opposition even in his lifetime (he died in 1916), and the idea of intelligent Martians was regarded as distinctly dubious. On the other hand, the idea that the dark areas were vegetation-tracts seemed plausible enough, and up to 1965 very few astronomers doubted it.

One piece of evidence, often quoted, was that of the so-called 'wave of darkening', bound up with the seasonal cycle of the polar caps. There is no doubt at all that the caps wax and wane according to the Martian seasons; during winter a cap is large, but shrinks rapidly with the oncoming of spring, and in midsummer it becomes very small indeed. (On some occasions, the southern cap has been known to vanish completely.) It was claimed that when a cap shrank, the dark areas near its border became sharper and more prominent, as though the lowly vegetation were being revived by the moisture-laden winds. Everything seemed to fit in, and most astronomers were confident that the overall picture of conditions on the Martian surface was not far from the truth.

I was always highly sceptical about the 'wave of darkening', because I was never able to follow it even though I had the advantage of observing Mars with large telescopes (including the Lowell 24-inch refractor at Flagstaff). On the other hand, I was in no real doubt that Mars was well able to support life of a kind, and it seemed probable that the grey patches were depressions, no doubt the beds of dried-up seas. The atmospheric pressure was thought to be around 85 millibars, equivalent to the density of the Earth's air at a height of between 50 000 and 60 000 feet above sea level, and it was tacitly assumed that the main atmospheric constituent was nitrogen.

Then, in 1965, came the flight of Mariner 4, and all these well-established views were rudely shattered. Pictures sent back from within a few thousand miles of the Martian surface showed that instead of being smooth, the planet was extremely rugged; there were large craters, together with mountains and valleys. Better pictures were obtained from Mariners 6 and 7, in 1969, but these were outshone by the results from Mariner 9, which was sent up in 1971. Instead of by-passing Mars, it was put into a closed path round the planet, and sent back thousands of close-range photographs.

The findings were dramatic. Mars is a world of giant volcanoes, huge craters, and spectacular valleys which look very much as though they have been water-cut. Some of the volcanoes, such as Olympus Mons, are 'shield volcanoes' of the Hawaiian type, but far loftier and more massive; Olympus Mons itself – recorded as a bright point by Schiaparelli, and named by him 'Nix Olympica', or the Olympic Snow – towers to a height of 15 miles above the outer country, so that it is three times the height of our Everest and dwarfs our most massive shield volcanoes, Mauna Loa and Mauna Kea in Hawaii. (Looking back in my notebook, I find that I recorded it as long ago as 1939, with a 6-inch refractor, though I had no idea what it was.) Equally interesting are the valleys, which seem to make up parts of drainage systems associated with the volcanoes. Such is the Valles Marineris, a tremendous formation with what seem to be tributaries extending from it. Of special interest is Hellas, the circular feature south of the V-shaped Syrtis Major. Instead of being a high plateau, it turns out to be the deepest depression on Mars. At times it is bright, and looks rather like an extra polar cap.

The Mariners proved, too, that the dark regions are not sunken; some, including the Syrtis Major, are raised. They are distinguished only by their colour, not by any marked difference in terrain, and they are not vegetation-covered. The canals were conspicuous only by their absence; they were simply due to tricks of the eye. Gone, too, was the whole idea of the 'wave of darkening'. And it was found that the atmosphere was thinner than had been expected, with a ground pressure of below ten millibars everywhere; the main constituent was not nitrogen, but carbon dioxide.

Next, in 1976, came the controlled landings of Vikings 1 and 2, in the ochre tracts of Chryse and Utopia respectively. Magnificent pictures were obtained of the rock-strewn surface, and the only real disappointment was the failure to detect any organic material, so that Mars may well be completely sterile. The orbiting sections of the

Vikings showed that the polar caps are made up of a mixture of water ice and solid carbon dioxide rather than being millimetre-thick layers of hoar-frost, as had been widely believed. There was plenty of evidence of past running water activity, and some of the features could hardly be anything other than old riverbeds. Today, the atmosphere pressure is too low for liquid water to exist, so that in the remote past the Martian atmosphere must have been much denser than it is now. Whether life there existed millions of years ago, and has now died out, is a question which we will not be able to answer until we can bring back the first specimens of Martian surface material.

Bad drawings of Mars are regrettably common, even in textbooks. It is very easy to 'see' what one expects to see, and for this reason I recommend going to the telescope with a completely open mind. Tables given in yearly almanacs can be used to work out the longitude of the central meridian for any particular time, but such calculations should be made after the observation rather than before.

Since drawings of Mars have to be made with comparatively high powers, the planet is a difficult object for small telescopes. A 3-inch refractor will show the polar caps and the main dark areas, such as the Syrtis Major and Acidalia Planitia, but for useful work at least a 6-inch is needed, preferably with an equatorial mount and clock drive. A scale of 2 inches to the planet's diameter is customary. When the phase is marked, as it will be unless the planet is near opposition, the disk should always be drawn to the correct shape. The outlines given here should be a help; I suggest photocopying them and using them as blanks (Fig. A6.2).

Begin, as always, by looking carefully at Mars for some times until your eyes have become thoroughly dark-adapted. Then sketch in the main details, such as the caps and the dark areas, using a moderate power. Change to the highest magnification which will give a sharp, crisp image, and fill in the finer detail. As soon as this has been done, note the time, and make a record of it below your sketch. This is important, because, as I have said, the drift of the markings across the disk becomes noticeable over quite short periods. (Obviously, any particular feature will pass over the central meridian of Mars about half an hour later each night, because the rotation period is half an hour longer than ours.) Finer details can then be added at leisure. Colours, intensities of the dark areas, and any clouds should always be looked for, together with features such as a dark border to the polar cap often seen when the cap is shrinking. Also, make full use of coloured filters, as with Venus.

Very small telescopes are useless for serious observations of Mars, but oddly enough it has been claimed that very large telescopes are also unsuitable. One famous textbook* states that 'if the aperture exceeds about 12 inches, the atmosphere will seldom allow the full aperture to be used'. This is a well-worn argument, but it is completely false. It is true that increased magnification will also increase any unsteadiness due to the Earth's air, but under even tolerable conditions a large telescope will always show more than a small one. This has been my own experience, with instruments ranging from a 3-inch refractor up to the Yerkes 40-inch, and I have never had the slightest temptation to stop the aperture down. However, a reflector of

---

* J.B. Sidgwick, *Observational Astronomy for Amateurs* (Faber and Faber, London). The book has run to many editions, and is still in print, though the original 1955 version remains the best.

from 8 to 12 inches is large enough for the amateur to play a part in Martian observation. Drawings made with smaller telescopes are bound to be rather suspect.

In view of what we have found out from the various space-probes, it may be asked: What can the amateur observer still do? The question is reasonable enough, but there is a firm answer to it. Mars, remember, has an atmosphere, and it is certainly not a static world. Dust-storms occur, and may at times cover the whole planet, hiding the surface detail completely; there are also well-defined clouds which shift and change, providing us with useful details about the Martian wind systems. There are the changing shapes and countours of the polar caps, and also irregular variations in the forms of the dark areas. Mars is never dull.

Finally there are the two tiny satellites, Phobos and Deimos, discovered in 1877 by the American astronomer Asaph Hall. Both are less than twenty miles across, so that even when Mars is near opposition they are hard to glimpse. I have seen both with my 15-inch reflector, and keener-eyed observers may catch sight of them with a 12-inch under ideal conditions, but not easily.

A rather stupid mistake on my part may show that it is never wise to reject an observation because it does not fit in with what is expected. I was once observing Mars with my 15-inch when I recorded a minute star like point, visible only when Mars itself was hidden by an occulting bar. I assumed that it was Phobos, but when I went to my tables I found that Phobos was nowhere near the position I had recorded, so that I dismissed the observation as being either an error or else due to a faint star. It was only on the following day that I found that the observation itself was quite correct; I had made a slip in my calculations.

Phobos is a peculiar little body. It whirls round Mars at a distance of only 3800 miles above the surface, about as far as London is from Aden, and has a revolution period of only about $7\frac{1}{2}$ hours. So far as Mars is concerned, the Phobos 'month' is shorter than the 'day', and to an observer on the planet Phobos would rise in the west, gallop across the sky in only $4\frac{1}{2}$ hours, and set in the east. For long periods it would be eclipsed by the shadow of Mars, and neither it nor the even smaller Deimos would be of much use as a source of moonlight.

Both satellites were photographed from Mariner 9 and the Vikings. Each is irregular in shape, and pitted with craters. Phobos and Deimos are quite unlike our Moon, and may well be ex-asteroids captured by Mars in the remote past. In 1988 the Russians launched two probes to Phobos, though unfortunately contact with the first of these was lost after a month or so. Phobos 2 entered an orbit round Mars in January 1989, but contact with it was lost before it had produced really valuable results.

Our knowledge of Mars has grown out of all recognition since 1965. Where we had hoped to find a life-bearing world, with vegetation tracts and a reasonably useful atmosphere, we have in fact found a planet which is hostile by our everyday standards. The massive shield volcanoes, the deep rift valleys and the craters present a scene which is quite unlike the Earth, and not really similar to the Moon. Yet Mars remains the most fascinating of all the worlds in the Solar System, and there is every hope that it will be reached within the next few decades. It will be left to Man to bring life at last to the Planet of War.

10

The Outer Planets

If you look at a scale map of the Solar System, you will see at once that the division of the planets into two main groups is very marked. Between the orbits of Mars and Jupiter there is a gulf over 300 million miles wide.

Nearly two centuries ago, Johann Elert Bode suggested that there might be a small planet moving round the Sun at a mean distance of about 260 million miles. There were sound reasons for believing that he might be right, and in 1800 a group of leading astronomers, headed by Schröter and the Baron Franz Xavier von Zach, began a systematic hunt for the missing body. Ironically, they were forestalled. Before the scheme was in full working order, the Italian astronomer Piazzi, at the Palermo Observatory, came upon a starlike object which turned out to be a small world circling the Sun at almost the correct distance. It was named Ceres, in honour of the patron goddess of Sicily.

Ceres was too faint to be seen with the naked eye, and it was also small; the modern value for its diameter is a mere 579 miles. It seemed too insignificant to rank as a true planet, and Schröter and his 'celestial police' went on with their programme. Between 1801 and 1808 they discovered three more minor planets – Pallas, Juno and Vesta – and when a fifth (Astræa) was added in 1845, long after the 'police' had disbanded, it became clear that the original four were only the brightest members of a whole swarm. Since 1848 no year has passed without fresh discoveries, and over 3000 of these minor planets or asteroids are now known. The total number has been estimated as at least 40000. It used to be thought that they represented the fragments of an old planet (or planets) which met with disaster in the early story of the Solar System, but most astronomers now think that the material of the asteroid belt never formed part of a larger body. Every time such a body tried to form in that particular region, the powerful gravitational pull of Jupiter disrupted it.

Ceres remains much the largest member of the swarm, and of the rest only a few have diameters of over 200 miles. Some are real midgets – less than a mile across – so that there is no real distinction between a small asteroid and a large meteoroid. Of all the asteroids only one, No. 4 (Vesta), can ever be seen without optical aid, though binoculars will show quite a number.

Hunting and photographing asteroids is a pleasant pastime, and it is not difficult. I once spent an evening searching for known asteroids; I was using a 6-inch refractor, and I observed 15 asteroids in only two hours, though they looked exactly like stars and I could not identify them all until I reobserved on the following night.

The procedure is to look up the position of the target asteroid, using the tables in a yearly almanac, and plot it on your star-chart. Then go to the telescope, and search until you are sure that you have found the star-field (obviously a low magnification is suitable). As the minor planet will appear exactly like a star, it will not be recognizable on sight, so the only course is to make a chart of all the stars in the area. When you look again, on the next clear night, the stars will be unchanged, but the minor planet will have shifted.

Though most of the asteroids keep strictly to the main zone between the orbits of Mars and Jupiter, some have unusual paths. The 'Trojans' have the same mean distance as Jupiter, so that they are very faint, while the extraordinary Hidalgo, only about 5 miles across, has an eccentric orbit which carries it from inside the path of Mars our almost as far as Saturn. And in 1977 Charles Kowal, at Palomar, discovered a relatively large asteroid, Chiron, whose orbit lies mainly between those of Saturn and Uranus. Chiron has a period of 50.7 years; when it next reaches perihelion, in 1995, its magnitude will rise to 15, so that amateurs equipped with comparatively large telescopes will be able to locate it. Its precise nature is uncertain.

On the sunward side, there are occasional asteroids which swing inward, and may make close approaches to the Earth. Eros, the first-known of them, was discovered as long ago as 1898; it can come within 15 million miles, and the magnitude is then only just below 8, so that it is within binocular range, but its longest diameter is no more than 20 miles — like many small asteroids, it is irregular in shape. Other 'Earth-grazers' can come even closer. Hermes, only a mile across, whirled by us in 1937 at only 480000 miles, less than twice the distance of the Moon. When the news was released, some people became alarmed at the idea of a celestial collision, and one national newspaper produced the immortal headline: 'World Disaster Missed by Six Hours as Tiny Planet Hurtles Earthward'. In 1989 an even smaller asteroid passed by at a mere 429000 miles. One, midget, Toro, has an orbit not too unlike that of the Earth, and when it was identified, in 1948, some press reports described it as a minor earth satellite! Oddest of all, perhaps, are Icarus and Phæthon, which have perihelion distances within the orbit of Mercury, so that when at their closest to the Sun they must be red-hot.

All of these 'Earth-grazers' are very small, so that few of them become visible in modest telescopes. It has been suggested that they are not genuine asteroids at all, but are ex-comets which have lost all their gas and dust. I am sceptical about this idea, but I admit that it cannot be ruled out.

Beyond the main asteroid zone we come to mighty Jupiter, giant of the Solar System. Though it never comes much within 360 million miles of us — well over a thousand times as remote as the Moon — Jupiter shines so brilliantly in our skies that it cannot possibly be mistaken for a star; it is outshone only by Venus, and very occasionally by Mars. Jupiter is also convenient inasmuch as it is always well seen for several months in every year.

Jupiter's vast globe could swallow up 1300 bodies the volume of the Earth, but it is not so massive as might be thought. If we could put Jupiter into one pan of a gigantic pair of scales, we would need only 318 Earths to balance it. This must mean that Jupiter is less dense than the Earth, and in fact the mean density is only 1.3 times that of water.

Jupiter is not a solid, rocky body. When you look at it through a telescope, what you see is not a hard surface, but a cloudy vista with details which change not only from night-to-night, but from hour-to-hour. Yet there is no comparison with the clouds of Venus. Jupiter's 'atmosphere' is so deep that it merges into the true 'body' of the planet. According to modern theory, there is a solid core made up of iron and silicates at high temperatures; around this is an extensive 'shell' of liquid, mainly hydrogen, above which comes the 600 mile deep layer of gas. Spectroscopic analysis of the upper gas shows it to be an unprepossessing mixture of ammonia, methane and free hydrogen, with a good deal of helium as well. The upper clouds are bitterly cold.

Four space-probes have now by-passed Jupiter: Pioneers 10 and 11, and Voyagers 1 and 2. They have found that there is a thin, dark ring, too elusive to be recorded from Earth; the magnetic field is extremely powerful; and the planet is surrounded by zones of radiation which would be instantly lethal to any astronaut foolish enough to venture inside them. The escape velocity is a full 37 miles per second – which is why Jupiter, unlike the Earth, has been able to hold on to its hydrogen.

In a small telescope Jupiter appears as a yellowish disk, flattened at the poles and crossed by prominent streaks known as belts. Increased power shows finer details such as spots, wisps and festoons (Fig. 10.1). Obviously all these are phenomena of the high atmosphere, but they change rapidly, and studies of them are of tremendous importance in our efforts to understand what Jupiter is really like. The main work has been done by amateurs, and the records of the Jupiter Section of the British Astronomical Association are probably the most complete in existence; they go back to 1890.

Apparently, gases warmed by the internal heat of Jupiter rise into the upper atmosphere and cool, forming clouds of ammonia crystals floating in gaseous hydrogen. The clouds form the bright zones on the disk, which are both colder and higher than the dark belts. Several of the belts and zones are relatively stable, changing only slightly in latitude, while others may disappear for a while. On most occasions the two main belts lie to either side of the Jovian equator, but the South Equatorial Belt is notoriously variable in both width and intensity. The strangest views I have ever obtained were in 1962–64, when the two equatorial belts ran together to form a single dark feature; in 1988 the South Equatorial Belt was broader than its northern counterpart, but in mid-1989 it almost vanished for a while.

Spots are common, and many of them are bright and white. Most of them are relatively short-lived. The chief exception to this rule is the Great Red Spot, which became striking in 1878 and can be traced on drawings dating back to the seventeenth century (Fig. 10.2). In its prime, the Spot was a brick-coloured elliptical object over 22000 miles long by 7000 miles wide; it is associated with a 'hollow' which intrudes

Fig. 10.1. Jupiter: 7 April 19, 2042 GMT, 419 mm reflector, × 248. Drawing by Paul Doherty, showing the shadows of Io (preceding or left) and Europa (following or right). Europa itself is shown off the preceding limb of Jupiter.

into the South Temperate Belt. It is not always on view, and has been known to vanish for a period of several years, but the Hollow can almost always be traced.

Before the space-probe era, the nature of the Red Spot was not known; theories ranged from a Jovian volcano (!) to the top of a column of stagnant gas, or even a solid body floating in the clouds. We now know that it is a whirling storm – a phenomenon of Jovian 'meteorology'; it is rotating, and the red colour may well be due to nothing more nor less than phosphorus. There is evidence that it is shrinking, and it may not last indefinitely, but we will almost certainly be able to follow it for many years yet. It is the only feature of its kind which has persisted for more than fifty years, its nearest rival being a disturbance in the south tropical zone which lasted from 1901 to 1940.

If Jupiter is watched for a few minutes with a magnification of 150 or more, the surface features will be seen to be drifting slowly across the disk. This is because of the planet's axial rotation, and is much more obvious than with Mars, since Jupiter spins much more quickly. It does not rotate in the way in which a solid body would do. In

Fig. 10.2. Photograph of Jupiter, showing the belts and the Great Red Spot. (Photograph by H.E. Dall, 12-inch reflector)

the equatorial zone, between the two main belts, the mean rotation period is 9 hours 50 minutes 30 seconds (System I). The rest of the planet has a mean period of 9 hours 55 minutes 41 seconds (System II). But these figures are only means; thus in System I there is a region of even more rapid rotation known as the Great Equatorial Current.

Moreover, individual features may have rotation periods of their own, so that they drift around in longitude. Between 1901 and 1940 the Red Spot and the South Tropical Disturbance were both to be seen in much the same latitude, and the Disturbance moved the more quickly of the two, so that periodically it caught up with the Spot and 'lapped' it; when the two were close together, there were obvious interactions. In 1919–20 and 1931–34 the British Astronomical Association observers even recorded 'circulating currents' in the South Tropical Zone. There is always something new to see on Jupiter. I well remember that in 1967 I discovered and followed a minor red spot in the South Tropical Zone which lasted for several weeks before fading away.

Jupiter's quick rotation means that you cannot afford to be too leisurely when making a disk drawing. The sketch should be completed in less than ten minutes, as otherwise the drift of the surface features will introduce errors. As with Mars, the main details should be filled in first; the time should be noted, after which the finer details can be added with a more powerful eyepiece.

One minor irritation is that you cannot use a pencil compass to draw the outline of the disk. The polar flattening amounts to some 6000 miles (as against only 26 miles for the solid, slower-spinning Earth), and it cannot be neglected. Shaping the outline freehand is a tedious business, at least for somebody who is as clumsy as I am. My method is to use prepared blanks. I have had mine printed, but there is no objection to making photocopies of the blank given in Appendix 7.

The rotation periods of specific features can be worked out by observing their transits across the central meridian. Visual estimates can be surprisingly accurate, and Jupiter spins so quickly that it is often possible to time 20 or 30 transits per hour. On several occasions during 1967, when Jupiter was well placed for British observers, I was able to observe continuously for more than a complete rotation, so that I was able to take successive transits of the Red Spot during the same session. It is hardly necessary to add that a reliable watch is essential — and make sure that it is set to the correct GMT.

Once the time of transit has been found, the longitude of the feature can be found by means of tables in a publication such as the *Handbook of the British Astronomical Association*. This is an easy process, and involves nothing more frightening than simple addition.

Transits assumed extra importance in 1955, when two American researchers, B.F. Burke and K.L. Franklin, found that Jupiter is a source of radio waves. The discovery was unexpected, and radio astronomers naturally wanted to know whether the emission came from the whole planet, or merely from small active regions such as the Red Spot. If the latter were true, then the radio emission would be strongest when the source lay near the central meridian. The amateur observers were called in, and it was eventually found that there is no marked correlation between radio strength and visual features, though there is a connection with the position in orbit of Jupiter's largest inner satellite, Io.

For routine work on Jupiter, a power of 150–250 on a 6-inch reflector is adequate. Transits can be taken almost as accurately as with a larger instrument, but there will be fewer observed, since the minor features will not be visible. Some years ago I did some careful checking with three other regular observers of Jupiter; we found that our transit times were always in agreement to within an error of half a minute.

Jupiter has an extensive family of satellites. Most of them are beyond the range of amateur-owned instruments, but four are bright enough to be seen with any small telescope; good binoculars will show them, and there are even a few keen-sighted people who have glimpsed one or two with the naked eye alone. They were first studied in detail by Galileo, in 1610, and are known collectively as the Galileans; their individual names, reckoning outward from Jupiter, are Io, Europa, Ganymede and Callisto.

Io is slightly larger than our Moon, Europa slightly smaller, and Ganymede and Callisto much larger; indeed, Ganymede is larger than the planet Mercury, though less dense and less massive. Their movements are fascinating to watch. Since all four move more or less in the plane of Jupiter's equator, they generally line up, but they are not hard to identify; Ganymede is always the brightest, and Callisto the faintest. Their positions for each night are given in astronomical yearbooks.

They are not always on view. A satellite may pass in front of Jupiter, appearing in transit; it may pass behind, and be occulted; it may pass into Jupiter's shadow, so that it is eclipsed. Shadow transits can be striking, and there are occasional mutual phenomena, when one satellite eclipses or occults another.

Surface details on the Galileans are almost impossible to see from Earth, even with giant telescopes, but thanks to the Pioneers and Voyagers we now have accurate maps of them. They are indeed remarkable bodies. Ganymede and Callisto are icy and cratered; Europe has a smooth surface which seems to be icy, and Io is red, with violently active sulphur volcanoes. Io is connected with Jupiter by a powerful magnetic current (which is why it has a marked effect on Jovian radio emissions), and since it moves in the midst of the radiation zones it may lay claim to being the most hostile world in the entire Solar System.

The remaining twelve satellites are extremely faint, and even the brightest of them, Amalthea, cannot be seen except with a telescope of more than 18-inches aperture, so that for the moment they need concern us no further. The outer four move round Jupiter in a retrograde direction, and may well be captured asteroids rather than *bona fide* satellites.

Far beyond Jupiter, at an average distance of 886 million miles from the Sun and a minimum of 741 million miles from the Earth, lies Saturn, second of the giant planets. It is smaller than Jupiter, with an equatorial diameter of just under 75 000 miles and a mass 95 times that of the Earth, but in composition it is essentially similar, though the density is less; in fact the mean density of Saturn is actually less than that of water. The synodic period is 378 days, so that Saturn comes to opposition about a fortnight later each year; the planet's own 'year' is 29.5 times longer than ours. The quick rotation, on average 10 hours 39 minutes, means that the globe is obviously flattened.

Saturn shows belts and spots, but the surface features are much less conspicuous than those of Jupiter, and well-marked spots are rare. The most spectacular outbreak that I have seen occurred in 1933, when a bright white spot near Saturn's equator was discovered by a most unusual amateur astronomer – W.T. Hay, better known to the public as Will Hay, the stage and screen comedian. It became prominent enough to be visible with the 3-inch refractor that I was using at the time, but it did not last for more than a few weeks. Features of this kind can be used for transit observations, as with Jupiter, but they are so unusual that a close watch should always be kept for them.

Saturn is a less active world than Jupiter, but the various zones show changes in brightness, so that intensity estimates are valuable. These can be made by eye estimates, on a scale of from 0 (brilliant white) to 10 (black shadow). The work needs a telescope of at least 8-inches aperture for a Newtonian reflector, but Saturn is a convenient object inasmuch as it will usually stand a comparatively high magnification.

The glory of Saturn lies in its ring system. Huygens, the leading telescopic observer

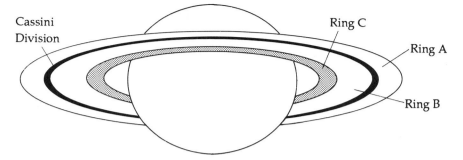

Fig. 10.3. Diagram of the main ring-system of Saturn.

of the seventeenth century, described it as 'a flat ring, which is inclined to the ecliptic and which nowhere touches the body of the planet', but actually there are three rings within the range of moderate telescopes, two bright and one dusky (Fig. 10.3). The whole system has a diameter of about 170 000 miles.

Saturn is a massive planet, and has a strong gravitational pull. Were the rings liquid or solid, they would soon be broken up and destroyed; for many years we have known that they are made up of small, icy particles whirling round the planet in the manner of miniature satellites. It is possible that they are the shattered remnants of a former large satellite which was broken up, though on the whole it seems more likely that they represent the débris left over, so to speak, when Saturn and its system were being formed.

A 3-inch telescope will show the rings; in a 6-inch the sight is glorious indeed, and Saturen is without doubt the most beautiful object in the entire sky.

Details can be seen in the ring-system when it is well placed. The two bright rings, A and B, are separated by a dark gap known as Cassini's Division in honour of its discoverer, the Italian astronomer who first reported it as long ago as 1675. A second gap, in Ring A, can be seen with an 8-inch telescope under good conditions; it is called Encke's Division. Other minor gaps, not definitely known before the flights of the Voyager probes, are beyond the range of most Earth-based telescopes, though some of them have been reported now and then. Ring C, the Crêpe or Dusky Ring, lies closer-in than the bright pair, and is semi-transparent. It is extremely hard to see unless the ring-system is wide open.

Though the rings are so extensive, measuring around 170 000 miles from one side to the other, they are also extremely thin (Fig. 10.4). When placed edgewise-on to the Earth, they almost disappear, as last happened in 1980 and will happen again in 1995. Both the Earth and the Sun pass through the ring-plane at such times; when the Sun does so, the rings are practically unilluminated. It is often said that they vanish completely at edgewise presentations. This is not quite true; on the last occasion I was always able to follow them as an excessively thin and faint line of light, but I had the advantage of using the Lowell 24-inch refractor at Flagstaff, and I agree that the rings would not have been seen with smaller instruments, though their invisibility does not last for long.

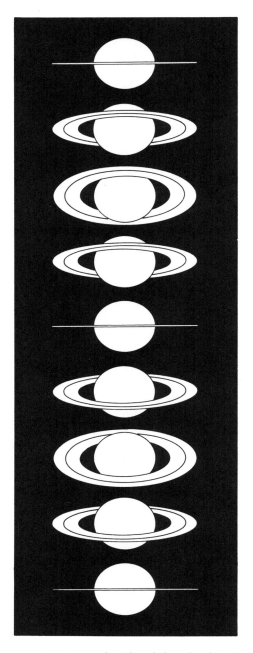

Fig. 10.4. Saturn's rings seen at various angles. The whole cycle takes 15–17 years.

Saturn is an awkward object to draw. Prepared outlines may be used, and the views here give the appearances at different stages of the ring-system, but there is no real short cut, as with Jupiter. For example, the Cassini Division looks broadest near the ring-tips or ansæ, and the Encke Division can never be seen all the way round. Points to note include the intensities of the various rings (B is always brighter than A), the shadow effects of rings on disk and disk on rings, and the visibility – or otherwise – of the Divisions. Occasionally Saturn occults a star, and these occultations are interesting, because even the bright rings may not be dense enough to hide the star completely. Incidentally, Ring C has been suspected of variations in brightness.

Most of our detailed knowledge of Saturn has been drawn from two space-probes; Voyager 1, which by-passed Saturn in November 1980, and its twin Voyager 2, which followed in August 1981. It was found that the rings are much more complicated than had been thought, and there are thousands of narrow ringlets and minor divisions; there are thin rings even inside the Cassini Division, and several extra rings were found outside the main system. (Ring D, reported by some Earth-based observers as another dusky ring closer-in than Ring C, does not exist *as* a ring, though there are considerable numbers of particles in that region.) The Voyager pictures showed abundant detail on Saturn's disk, and it was also found that there is a magnetic field, though the radiation zones are not nearly so powerful or as lethal as those of Jupiter.

Before the Voyager missions, Saturn was known to have at least 10 satellites, but the family is quite unlike Jupiter's. Whereas Jupiter has four large satellites and twelve very small ones, Saturn has one major satellite (Titan), four of moderate size (Rhea, Iapetus, Dione and Tethys), three which are considerably smaller again (Hyperion, Enceladus and Mimas), and several midgets, of which the outermost, Phœbe, moves round Saturn in a retrograde orbit at a distance of 8 000 000 miles, and may well be a captured asteroid. With my 15-inch reflector I have seen eight of the satellites, and a 6-inch should show the five senior members of the family, though only Titan is really bright.

Titan, in fact, is one of the most interesting bodies in the Solar System. Before the Space Age it was already known to have an atmosphere, but we did not know which gases were present – methane and carbon dioxide were two favoured candidates – and neither did we know how thick it was, or whether it was cloudy. I was in Mission Control at Pasadena, California, when Voyager 1 closed in on Saturn, and I remember an argument with my colleague Dr Garry Hunt. I thought that we would see Titan's surface through the clouds. He disagreed. In the event, he was right and I was wrong. All that Voyager showed us was the top of a layer of orange cloud. The real surprise came when it was found that the main constitutent is not carbon dioxide, but nitrogen, though admittedly there is plenty of methane as well. The ground pressure is about 1.5 times that of the Earth's air at sea-level. We still do not know what the surface of Titan is like; there may be seas of liquid ethane, cliffs of solid methane, and a methane rain dripping down all the time from the orange clouds in the nitrogen sky. All the basic ingredients for life exist there, though it seems that the bitterly cold climate must have prevented life from gaining a foothold. Unfortunately we are not likely to find out more until the dispatch of another space-craft, and as yet we cannot tell when that

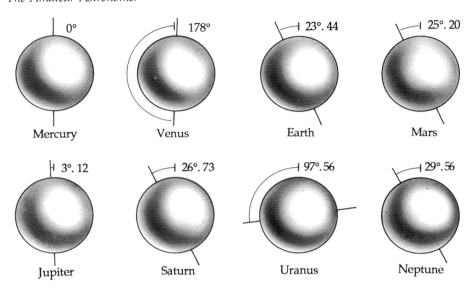

Fig. 10.5. *Axial inclinations of the planets. Obviously, the tilt of Uranus is exceptional, as it is more than a right angle to the perpendicular to the planet's orbit.*

will be. (Voyager 2, incidentally, did not survey Titan from close range. Had it done so, it would have been unable to go on to rendezvous with its two final targets, Uranus and Neptune.)

Iapetus and Rhea can be seen with a 3-inch refractor. Iapetus is of special interest, because it is so variable in light; when west of the planet it is conspicuous enough, but becomes much dimmer when to the east of Saturn. Voyager pictures show that it has one hemisphere which is dark and one which is bright. Iapetus takes 79 days to complete one journey round Saturn, and this is also the period of its axial rotation, so that during western elongations it is the bright side which is turned toward the Earth. The Voyager results show that most of the satellites, apart from Titan, are icy and crater-scarred, though for some unknown reason Enceladus — like Europa in Jupiter's system — has large areas which are more or less smooth.

Two of the smaller, closer-in satellites — Janus and Epimetheus — share an orbit, and are almost certainly the fragments of a former larger satellite which broke up. I have to admit that I overlooked Janus completely in 1966, when the rings were edgewise-on and I was making observations with the 10-inch refractor at the Armagh Observatory. Janus was reported by the French astronomer Audouin Dollfus, from the Pic du Midi. When I checked my own work, I found that I had recorded Janus on at least four occasions — but since I did not realize that it was new, I can claim absolutely no credit!

Phenomena of the larger satellites can be followed with a telescope of 8 inches aperture or more, but they are none too easy to see except in the case of Titan. Occultations of the satellites by the rings can also be seen now and then.

Far beyond Saturn we come to the next giant planet, Uranus, which is just visible with the naked eye when you know where to look for it. In spite of its great distance,

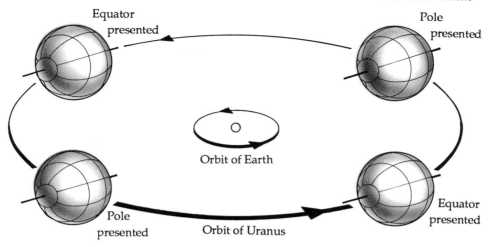

Fig. 10.6. The changing presentation of Uranus. Sometimes (as in 1986) a pole is presented to the Earth and the Sun; at other times the equator is presented.

never less than 1600 million miles from the Earth, a small telescope will show that it is not a star; it presents a pale, greenish disk, though markings on it are almost non-existent (I have seen a few brightish and occasional duskier regions, but not with any certainty). The diameter is just over 30 000 miles, so that Uranus is much larger than the Earth, though its diameter is less than half that of Saturn.

Uranus is a celestial oddity. Whereas most of the planets have their axes of rotation only moderately inclined to the perpendiculars to their orbits (23½ degrees in the case of the Earth), Uranus has an axial inclination of more than a right angle, as shown in Fig. 10.5. Consequently, the 'seasons' there must be most peculiar. First much of the northern hemisphere, then much of the southern, is plunged into darkness for a period equal to 21 Earth-years, with opposite conditions in the other hemisphere. Sometimes we look straight 'down' on to a pole, as in the mid-1980s, while at other times it is the equator which is presented (Fig. 10.6). Nobody knows why Uranus spins in this extraordinary way; all sorts of theories have been put forward, but none can be said to be really plausible. The Uranian 'year' is 84 times as long as ours, so that the planet is a slow mover among the constellations. The rotation period is 17.2 hours.

In constitution, Uranus is rather different from Jupiter or Saturn. There is probably a solid core; much of the planet is made up of hydrogen and helium, while there is also a good deal of water, plus methane and other substances. Unlike the other giants, Uranus appears to lack an internal heat-source. The planet was surveyed by Voyager 2 in January 1986, but even from close range not many definite surface features were on view. There is a magnetic field, but the magnetic axis is inclined to the rotational axis by around 60 degrees, and does not even pass through the centre of the globe.

So far as the amateur is concerned, two kinds of research may be attempted. First, make regular estimates of Uranus' magnitude, which seems to be rather variable —

partly because of changes in the Sun's output, and partly, it is believed, because of changes on the surface of the cloud-layer itself. A low power is best for this work, because with higher magnifications Uranus appears as a definite disk, and is difficult to compare with a star. In 1955, when Uranus and Jupiter lay close together in the sky, I tried to compare Uranus with Ganymede and Callisto, but with scant success, because the planet looked much larger than the satellites and had a lower surface brightness. Secondly, there are the rare occasions when uranus occults a star. It was an observation of this kind, in 1977, which led to the original detection of Uranus' ring-system, because the occulted star 'winked' regularly both before and after it passed behind the disk of the planet. It was only with the Voyager 2 mission that we had our first detailed views of the rings. Unlike the glorious rings of Saturn, the system of Uranus is narrow and dark, so that it is hopelessly beyond the range of most Earth-based telescopes.

Before the Voyager pass, Uranus was known to have five satellites, all smaller than our Moon. Of these, Titania and Oberon are within the range of an 8-inch telescope, and perhaps Ariel; Umbriel requires at least a 12-inch, and Miranda is more elusive still (it was discovered, by G.P. Kuiper, as recently as 1948). Voyager showed that they, too, are icy and cratered, while Miranda shows a bewilderingly varied landscape unlike anything else we know. Ten smaller, closer-in satellites were discovered from Voyager, but as yet they have not been recorded by any Earth-based telescope.

The last of the giant planets, Neptune, moves round the Sun at the immense distance of 2793 million miles in a period of 164.8 years. Its magnitude does not rise above 7.7, so that it is well below naked-eye visibility; binoculars will show it, but a telescope is needed to resolve its tiny, bluish disk. In some ways Neptune is not unlike Uranus — it is very slightly smaller, but appreciably more massive — but it has an internal heat-source, and does not share Uranus' exceptional axial tilt.

Neptune was by-passed by Voyager 2 in August 1989, and found to be much more dynamic than the bland Uranus. The main surface feature is the Great Dark Spot, in the southern hemisphere, which is about the size of the Earth; above it, at a height of around thirty miles, are 'cirrus-type' clouds made up of methane ice. The Great Dark Spot has a rotation period slightly smaller than the mean length of the Neptunian 'day' (16 hours 3 minutes) and is periodically 'lapped' by a smaller dark spot, further to the south, and a cloudlike feature nicknamed the Scooter because of its faster rotation. Neptune's magnetic field is weaker than that of Uranus, but it too is inclined to the axis of rotation — in this case by about 50 degrees — and is offset with respect to the centre of the globe.

Before the Voyager mission Neptune was known to have two satellites, one large (Triton) and the other small (Nereid). Triton was unique among large satellites by having retrograde motion, while the orbit of Nereid was exceptionally eccentric. Voyager showed Triton to be a remarkable world covered with nitrogen and methane ice, possibly with active nitrogen geysers; six small inner satellites were also found.

Occultation methods had led to the suspicion that Neptune might have incomplete rings. Voyager showed that there are in fact three complete rings, of which there are brighter sections, but all are very faint.

It cannot honestly be said that much amateur work can be done with respect to Neptune unless the planet occults a star — and this does not happen often. Surface details, even the Great Dark Spot, are out of amateur range, and so are all the satellites apart from Triton, which is however brighter than any of the satellites of Uranus.

Neptune was discovered in 1846, by Johann Galle and Heinrich D'Arrest at the Berlin Observatory, following calculations made of the movements of Uranus. On the same basis, yet another planet was predicted by Percival Lowell (of Martian canal fame), and in 1930 Pluto was found not far from the position which Lowell had given. The discovery was made by Clyde Tombaugh, then a young amateur who had been called to Flagstaff specifically to take part in the search. Tombaugh, I am glad to say, is still very much alive, and is one of America's most senior and respected astronomers. It was a great honour for me to be invited to collaborate with him in writing the official book about Pluto.

Pluto has set astronomers problem after problem. It has an eccentric orbit, and for part of its 248-year period it is actually closer-in than Neptune, though its orbit is tilted to the ecliptic by the unusually sharp angle of 17 degrees, and there is no fear of a collision on the line. Pluto last reached perihelion in 1989, so that between 1979 and 1999 it temporarily forfeits its title of 'the outermost planet'. Even so, the magnitude never rises much above 14, so that it is beyond the range of small telescopes. A minimum of at least 8 inches aperture is needed.

The most curious fact about Pluto is its size — or lack of it. The diameter is now known to be no more than 1519 miles — considerably smaller than our Moon and also smaller than Triton. Apparently the globe is a mixture of rock and ice, and slight variations in brightness have shown that the axial rotation period is 6.3 days. It is not a solitary traveller; it has a satellite, Charon, with a diameter of around 745 miles. The orbital period is the same as the axial rotation period of Pluto, so that the two are 'locked'.

Amateur work is restricted to checking on possible occultations. Obviously these do not happen often, and I have never seen one, though I have recorded a couple of 'near misses'. I have already referred to the 1988 occultation which provided such valuable information about Pluto's atmosphere.

Because Pluto is so small, and so lacking in mass, it could not possibly exert any measurable effects upon the movements of giants such as Uranus or Neptune. Either Lowell's reasonably correct forecast was sheer luck, or else the real Planet Ten awaits discovery. I am convinced that it exists, but we have no real idea of where it may be in the sky, and it is bound to be extremely faint.

I hope I have said enough to show that in spite of all the space-craft results, coupled with the work being carried on at professional observatories, there is still a great deal of work for the amateur to do. He may not have a large telescope; he may not possess a science degree, but at least he can make himself useful. And after a lifetime's work, he will realize that there is still much that has been left undone.

11

Comets and Meteors

A brilliant comet, with a tail stretching half-way across the sky, must be one of Nature's grandest spectacles. Small wonder that it caused fear and panic in ancient times, when comets were believed to be heralds of disaster. Shakespeare wrote in *Julius Cæsar*:

> When beggars die, there are no comets seen:
> The heavens themselves blaze forth the death of princes,

and even today the feeling is not entirely dead. Many people will recall the end-of-the-world forecasts made at the time of the last return of Halley's Comet, even though the comet itself was never bright.

Broadly speaking, a comet is made of an icy nucleus only a few miles in diameter. When a long way from the Sun, there is little or no activity; but as the comet draws in, the ices become heated, and the comet may develop a tail or tails (Fig. 11.1). Yet the total mass is very slight by planetary standards, and even a direct collision with a cometary nucleus would do no more than local damage.

Comets are still sometimes confused with meteors, or shooting-stars. There is a close link between the two, since meteors are nothing more nor less than cometary débris; but whereas a shooting-star will dash across the sky and vanish in a second or two, a comet may remain on view for many months, moving so slowly against the starry background that its shift cannot be detected except over a period of hours. If you see something moving perceptibly, it certainly cannot be a comet. Also, by no means all comets develop tails. I well remember showing a telescopic comet to a friend of mine who knew little about astronomy and cared less. His comment was that it looked 'like a tiny patch of cotton-wool', and I could not disagree.

Note, too, that the gas in a comet's head and tail is immensely rarefied, and is millions of times less dense than the air that you and I are breathing. I once described a comet as being 'the nearest approach to nothing than can still be anything', and even a brilliant visitor is not nearly so important as it may look.

The tails of comets always point more or less away from the Sun, and are of two kinds (Fig. 11.2). The 'dust' tail is curved, and is due to tiny particles being pushed away by the pressure of sunlight. The gas or 'ion' tail is straighter, and is caused by the

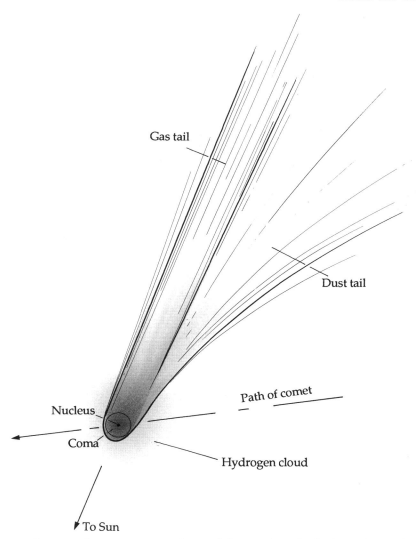

Fig. 11.1. Structure of a comet: coma, nucleus, and the two types of tails; the gas (ion) tail and the curved dust tail.

so-called solar wind, made up of electrified particles streaming outward from the Sun in all directions. The material making up the head ('coma') and tail is made up of material evaporated from the nucleus, so that each time a comet passes close to the Sun it wastes away; over a sufficient period of time all the volatiles will be lost. In fact, comets are relatively short-lived members of the Solar System.

Most of them move in very elliptical orbits, not in the least like those of the planets. The so-called periodical comets pass through perihelion in periods ranging from 3.3 years (Encke's Comet) to around 150 years; since they depend upon reflected sunlight, they can be seen only when they are moving in the inner part of the Solar System. (It is true that the cometary material can give out a certain amount of light on its own

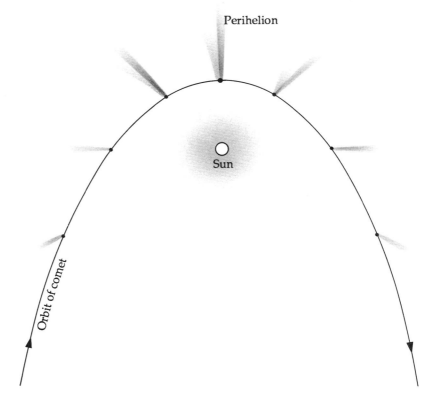

Fig. 11.2. Changing direction of a comet's tail. The tail always points more or less away from the Sun.

account, but the Sun is wholly responsible.) Several dozens of periodical comets have their aphelion points close to the orbit of Jupiter, and are known as Jupiter's comet family, though there is no likelihood that they were formed by being ejected from Jupiter itself.

Because of the continuous wastage, most of the periodical comets remain too faint to be seen without optical aid, though a few of them may just reach naked-eye visibility when best placed. The exception, of course, is Halley's Comet, which has a mean period of 76 years, and has been seen regularly ever since well before the time of Christ. It was seen in 1682 by Edmond Halley, later to become the second Astronomer Royal, and it was he who realized that it was identical with comets seen previously in 1607 and in 1531. He predicted that it would return once more in 1758, and it duly did so, though Halley himself did not live to see it. Since then it has been back in 1910 and in 1986.

The orbit of Halley's Comet is shown in Fig. 11.3. It reached aphelion in 1948, when it was well beyond the orbit of Neptune, and then started to draw inward, picking up speed as it moved along. It was recovered in 1982, by astronomers at the Palomar Observatory in California, and came within the range of amateur-owned telescopes in

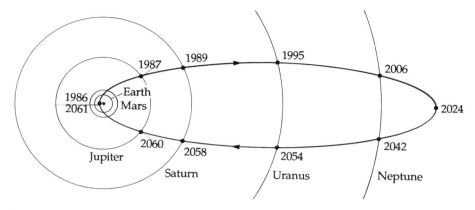

Fig. 11.3. Orbit of Halley's Comet. Perihelion was passed in 1986; the next will be in 2061. At present the comet is moving outward to aphelion.

mid-1985, by which time the tail structure was starting to develop. Sadly, the Earth and the comet were in the wrong places at the wrong times; at perihelion, in February 1986, Halley was on the far side of the Sun, and could not be seen at all. Even at its best, in March and April, it was still inconspicuous, though it was easy enough to see with the naked eye. It was then in the southern hemisphere of the sky; I had my best views of it from Australia.

To members of the general public, Halley's Comet was frankly disappointing, despite all the publicity it had received;* it was not nearly so bright as it had been in 1910. Yet scientifically it was of immense interest, and no less than five space-probes were sent to it: two Russian, two Japanese and one European. The European space-craft, named Giotto in honour of the Italian artist who had painted the comet in 1304, actually passed through the comet's head, and obtained close-range pictures of the nucleus. The results were in some ways unexpected. The nucleus was shaped rather like a peanut, nine miles long by five miles wide; it was made up of 'dirty ice', protected by an insulating layer of very black dust. It was also found that the dust-layer cracks when strongly heated by the Sun, so that the underlying ice is exposed; jets and streamers of gas and dust shoot out from the fissures, giving rise to all the activity of the comet. By now, 1989, the comet has started to move away to the further reaches of the Solar System, though it can still be followed with powerful telescopes; activity is dying down, and before long all that will be left will be the small, peanut-shaped nucleus. The next return will be that of 2061, though I am afraid that conditions will again be unfavourable. Things were very different at the return of 837 AD, when, according to contemporary records, the comet was brilliant enough to cast shadows.

It has been estimated that each time it passes through perihelion, Halley's Comet

---

* I cannot resist mentioning the Halley's Comet Society, founded by Brian Harpur, which has no aims, no objects, no rules and no ambitions; all it does is to meet occasionally on licensed premises. I always say that it is the only completely useless society in the whole of the world, apart, of course, from the United Nations. I am proud to be a member!

loses about 250 million tons of material. Obviously it cannot last indefinitely, and within a couple of hundred thousand years all its volatiles will have been exhausted.

Apart from Halley's, all the periodical comets are dim, with tails which are at best very faint (Fig. 11.4). Some of them have unusual orbits, a good example being Comet Schwassmann–Wachmann I, where the path is relatively circular and lies between the orbits of Jupiter and Saturn. It is usually very faint indeed, but sometimes suffers outbursts, brightening up sufficiently to come within the range of telescopes of modest aperture. I have observed it clearly with my $12\frac{1}{2}$-inch reflector.

Since several new comets are found each year, there must be some system of nomenclature. There are two systems, one provisional and the other permanent. First, the year's comets are allotted letters in order of discovery (a, b, c, etc); secondly, a comet is given a Roman numeral according to its order in passing perihelion. A comet which was, say, the second to be discovered in 1988 and was the fourth to pass perihelion in that year would become first 1988 b and then 1988 IV. Of course, there is no guarantee that a comet will reach perihelion in the year of its discovery; 1990 m or n may become 1991 II or III.

In most cases the names of the discoverers are used; two or three independent discoverers may be bracketed together, as with Schwassmann and Wachmann. Less often, the comet is named after the mathematician who first worked out its orbit. Halley's is the classic case of this, and Encke's is another.

Not all comets are seen regularly. In fact, all the 'great' comets which have been seen over the centuries have periods of hundreds, thousands or even millions of years, so that they cannot be predicted, and we never know when to expect them. They were fairly common in the last century, and there were some truly magnificent specimens. The comets of 1811 and 1843 had tails which reached right across the sky, while Donati's Comet of 1858 had a glorious triple tail. We will not see these comets again – at least, not in our time. Really brilliant comets have been rare of late, and the last was seen in 1910, a few weeks before Halley's.

However, there have been some comets which have become bright enough to attract general attention. Such was the Arend–Roland Comet of 1957, which showed what appeared to be a 'reverse tail' pointing sunward; actually it was not a tail at all, but merely débris which was spread along the comet's orbit and happened to catch the sunlight at a favourable angle. The Ikeya–Seki Comet of 1965 was brilliant for a few nights as seen from the southern hemisphere, and others have been Bennett's Comet of 1970 and West's of 1976. Some people will also remember Kohoutek's Comet of 1973, which was expected to become spectacular, but failed to do so. It was never very easy to see with the naked eye, but it was scientifically important; it was studied by the last crew of America's Skylab space-station, and was found to be surrounded by a vast cloud of tenuous hydrogen. Its estimated period is 75 000 years.

Whence come the comets? According to the theory first put forward by the Dutch astronomer Jan Oort, an immense swarm of comets exists at a distance of at least one light-year from the Sun. They are much too faint to be seen, but if a member of the 'Oort Cloud' is perturbed for any reason it may start to swing inward toward the Sun, finally coming within range of our telescopes. It may simply pass through perihelion

124

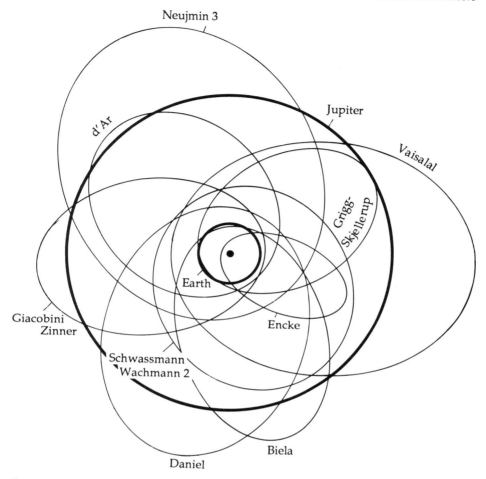

Fig. 11.4. Orbits of some periodical comets. The orbits of the Earth and Jupiter are shown for comparison.

and return to the Oort Cloud; alternatively it may be 'captured' by the gravitational pull of a planet (usually Jupiter) and forced into a short-period orbit, or it may even be put into an open orbit, so that it will be expelled from the Solar System altogether. At any rate, comets seem to be very primitive objects, dating back to the early story of the Sun's family.

When a periodical comet is due to return, its expected position is given in astronomical almanacs, and this is usually good enough for quick identification. Last time Encke's Comet came round I picked it up with my 3-inch refractor as soon as it came within range; and I am not, and never have been, a regular observer of comets.*

---

* Encke's Comet has the shortest known period. Its orbit is eccentric, but the aphelion distance is relatively small, so that powerful telescopes can now follow it all the way round its orbit – something which is not possible with most other periodical comets. Even Halley's is out of view for most of the time.

Comet-hunting is a fascinating occupation, and amateurs have a fine record; from Australia, William Bradfield has made no less then 13 discoveries. On the other hand, the observer must resign himself to many disappointments. He may not find a new comet for years; indeed, he may never find one at all.

Never use a high magnification. What is needed is a wide field of view, and in any case a powerful eyepiece is not of much use when studying a fuzzy, badly-defined object. Binoculars can be used to advantage, provided that they are of sufficient aperture. George Alcock, one of Britain's most successful hunters, uses binoculars which have 6-inch object-glasses.

Having selected the region to be swept, the telescope is moved slowly around in azimuth – and it may be added that for this sort of work, an ordinary altazimuth stand is as good as an equatorial. Stars, clusters and other objects will creep through the field, and if you relax your attention for even a moment you may miss a vital comet. (In other words – when comet-sweeping, don't blink!) At the end of the sweep the telescope is raised or lowered very slightly, and an overlapping sweep taken in the opposite direction. After you have covered the whole area in this way, repeat the whole performance until you are satisfied that you have overlooked nothing.

Immense patience is needed, and things are not helped by the presence of star-clusters and nebulæ, some of which look exactly like comets. *If you come across a misty object that is certainly not a star, do not jump to any hasty conclusions. When you look at your star-maps, you will probably find that your 'comet' is a cluster or a nebula that has been known for centuries (Fig. 11.5).

There is an interesting anecdote about this. Charles Messier, an enthusiastic comet-hunter of the eighteenth century, was persistently misled by uncharted stellar objects, and eventually he drew up a catalogue of them as 'objects to avoid', rather as a sea navigator charts rocks and shallows in a strait. Nowadays Messier's comets are all but forgotten, but we still use his catalogue of clusters and nebulæ.

Many comets will lie somewhere near the Sun's line of sight when they are drawing in toward perihelion, and most remain undiscovered until they are within the orbit of Mars, though of course there are exceptions to this rule. Also, the average comet brightens up considerably as it nears the Sun and the heat acts upon the ices of the nucleus. Therefore, the most promising areas for sweeping, for an observer in the northern hemisphere, are the west and north-west after sunset and the east and north-east before sunrise. It is also worth sweeping in the low north. There is no point in beginning until the sky is really dark, since a faint comet will be drowned by any background light.

Though more new comets will be seen in these directions than in others, there is no hard and fast rule. A comet may appear at any moment from any direction; it may have an open or a closed orbit; the orbit may be highly inclined, and it may have retrograde motion (as, indeed, Halley's Comet does). Flimsy, harmless and of negligible mass, comets can do no harm to anybody. And as I have commented, they are short-lived on

---

* In April 1986 I was in Australia, observing Halley's Comet. At that time it lay not far from the globular cluster Omega Centauri, and with binoculars or the naked eye the two looked very similar. I lost count of the number of people who came up to me claiming that they could see two comets instead of only one!

Fig. 11.5. *Halley's Comet near the Pleiades, 17 November 1985. I took this picture with a 210 mm zoom lens, f/4, exposure 20 minutes.*

the astronomical time-scale – and several short-period comets which used to return regularly have now vanished for good. Such are – or were – the comets of Biela, Westphal and Brorsen.

Apart from sweeping, the amateur who has equipped himself with an equatorial mounting, a measuring device, and perhaps a camera, can do valuable work in checking the positions and magnitudes of known comets – though it is not easy to compare an ill-defined comet with a star, which is to all intents and purposes a point source of light. Mathematically-minded enthusiasts may also care to try their hand at computing comet orbits.

The link between comets and meteors is shown clearly by the interesting case of Biela's Comet, whose peculiar career ended in spectacular fashion. The comet was discovered by Biela, an Austrian amateur astronomer, in 1826, and was found to be identical with comets previously seen in 1772 and in 1805. It was one of Jupiter's short-period group, and had a period of about 6.6 years. In 1826 it reached the third magnitude, so that it was quite conspicuous with the naked eye. It returned in 1832 as predicted, was missed in 1839 because it was badly placed in the sky, and returned on schedule in 1845.

Up to then, it had behaved in a perfectly normal manner, but during the return of 1845–46 it caused an astronomical sensation by splitting in half. Where there had been a single comet, twins could be seen, sometimes connected by a sort of filmy bridge. Sometimes the two were nearly equal, while at other times the original comet was the brighter. Both faded gradually into the distance, and the return of 1852 was eagerly awaited. This time the two comets were further apart, but both were quite distinct. For the 1859 return conditions were again hopelessly bad, but in 1865—66 the pair should have been easy objects. Yet they did not appear; so far as could be made out, Biela's Comet had disappeared from the Solar System. Comets have been nicknamed 'the ghosts of space', but no ghost could possibly have done a more successful vanishing act.

The next return should have taken place in 1872. Again the comet was absent, but a rich shower of meteors came from the position in the sky where Biela ought to have been. Coincidence could be ruled out, and for years afterwards meteors were seen each year at the time when the Earth crossed the path of the dead comet. Gradually the shower became feebler, and by now it seems to have ceased altogether, so that we have really seen the last of Biela's Comet.

It would be misleading to say simply that the comet 'broke up' into meteors, but the whole episode confirmed the view that meteors are simply cometary débris spread along the orbits. Most meteor showers, though not all, are identified with their parent comets — the Lyrids and the Eta Aquarids with Halley's Comet, the Taurids with Encke's Comet and so on.

Most people have wild ideas about the sizes of the particles which become incandescent in the upper air, and produce shooting-stars. Actually, the particles are very small indeed — most are no larger than grains of sand; a particle the size of a grape would produce a brilliant fireball. What we are seeing is not the meteor itself, but the luminous effects produced as it plunges to destruction, ending its journey to the ground in the form of fine dust. They become visible at heights of around 120 miles, and burn out before they have dropped much below 50 miles; their speeds may be as high as 45 miles per second relative to the Earth. Millions of them enter the upper atmosphere every day. There are also the so-called micro-meteorites, which have diameters of around $\frac{5}{1000}$ of an inch, and are too small to cause luminous effects.

Meteors travel round the Sun in elliptical orbits, sometimes as members of shoals (shower meteors) or on their own (sporadic meteors). Sporadic meteors may appear from any direction at any moment, but shower meteors are more obliging. If the Earth passes through a swarm, the ordinary laws of perspective will make the meteors appear to radiate from one particular point in the sky — just as the parallel lanes of a motorway will seem to radiate from a point near the horizon when you observe them from an overlooking bridge. This is shown in Fig. 11.6. The paths are really parallel, but seem to issue from the radiant point in the far distance.

A meteor shower is named according to the constellation which contains the radiant; the November Leonids from Leo (the Lion), the December Geminids from Gemini (the Twins) and so on. (The January Quadrantids come from an area which is marked by the now-rejected constellation of Quadrans, the Quadrant; the stars of the old Quadrans are now included in Boötes, the Herdsman.) Of course, this does not mean that all the meteors appear in the constellation which contains the radiant, but merely that they converge from that particular position. There are many annual showers, some richer than others; the most reliable shower is that of August, radiating from Perseus. The Perseids never fail us, and if you stare up into a dark, clear sky for a

Fig. 11.6. Principle of the radiant. I took this photograph from a bridge overlooking the M6 motorway; the parallel lanes seem to diverge from a point near the horizon.

few minutes at any time between the end of July and the beginning of the third week of August you will be unlucky not to see at least one shooting-star.

The richness of a shower is given by its ZHR, or Zenithal Hourly Rate. This is given by the number of naked-eye meteors which would be expected to be seen by an observer under ideal conditions, with the radiant at the zenith or overhead point. In practice, of course, these conditions are never fulfilled, so that the actual rate will always be less than the theoretical ZHR, but the values do give a good general indication. The Perseids usually have a ZHR of between 60 and 70, though of course it varies somewhat from one year to another.

Incidentally, the parent comet of the Perseids is Swift-Tuttle, which was discovered in 1862; it reached the second magnitude, with a 10-degree tail. The period was calculated at 120 years, but the comet has not been seen again; either the mathematicians have made a mistake, or else the comet has come and gone unseen.

Very occasionally we have what is termed a 'meteor storm', when the ZHR rockets up to many thousands for a few hours. This brings me on to the story of the November Leonids, which make up the most spectacular and the most erratic of all showers.

The parent comet, Tempel-Tuttle, has a period of 33 years. It was discovered in 1865, when it was of the fourth magnitude, and was identified with comets previously seen in 1366 and 1699. The meteors are 'bunched up' instead of being spread all round the comet's orbit, as with the Perseids, so that the best Leonid displays occur only every 33 years, when the comet is close-in. Records of them go back for many centuries. There were brilliant displays in 1799 and in 1833; equally good was the meteor storm of 1866, when it was said that meteors seemed to 'rain down from the sky like snowflakes' for a period of four hours. After 1866, the next display was due in 1899, but by then the swarm had been affected by planetary perturbations, and the main cluster missed the Earth, so that the expected shower did not materialize. The succeeding return was that of 1933 (not 1932). Again nothing of note was seen, but conditions seemed much more promising in 1966, and meteor observers were very much alert for the night of November 16–17. The comet itself was back, and came to perihelion in April of the previous year, though it was by no means conspicuous.

From Europe, the 1966 Leonids were most disappointing, and only a few were seen. Things were better elsewhere; as seen from parts of the United States (Arizona, for instance) the ZHR reached 100 000, making it the greatest display of the century. Maximum occurred at about 12 hours GMT, while it was daylight over Europe. In Britain, we missed the main shower by about six hours.

Since then, some Leonids have been seen every year during the nights before and after November 17. There is every chance that there will be another spectacular display in 1999, though I hasten to add that it is impossible to be sure.

To determine the velocity, height and orbit of a meteor, three types of data must be provided: the point of appearance of the meteor, the point of disappearance, and the duration (Fig. 11.7). Clearly it is necessary for the same meteor to be observed by two watchers at least twenty miles apart (more if possible), and all serious meteor observers work as members of teams.

No instruments are needed for meteor recording, but the observer has to have a really good knowledge of the constellations, as otherwise he will be unable to plot the track. The track should be plotted on a star map, but it is unwise to look down as soon as the meteor has vanished and try to remember where it went, because errors are bound to creep in. The solution is to check the path by holding up a rod or stick along the track which the meteor has followed, giving you the chance to take stock of the background and ensure that no mistake has been made. When you are satisfied, draw in the path on your chart and note the exact positions of the beginning and end of the track, and then write down: time of start, duration, duration of luminous trail, brightness (compared with a known star or stars), colour (if any), and any special features.

Meteor-watching is a lengthy and often a cold business. Standing or (preferably) sitting outdoors for hours on a chill winter night is enough to blunt the enthusiasm of the hardiest observer. Nevertheless, until recently most results were based upon the patient work of amateurs. It is fair to say 'until recently', because by now a new method of recording has been brought into operation – radar.

The passage of a meteor through the atmosphere has a pronounced effect upon the air-particles, and these effects can be detected by radar. Reduced to its barest terms, radar involves sending out an energy wave, and recording the echo as the wave is bounced back after hitting a solid object. Of course, a meteor trail is not solid, but it acts in much the same way, and radar detection of shooting-stars has now come into general use. It is unhampered by clouds or daylight, and it would be idle to pretend

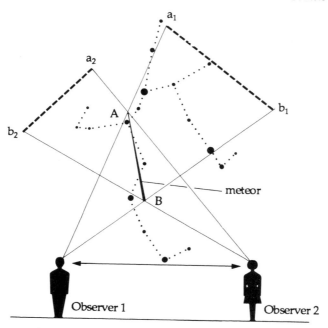

Fig. 11.7. Measuring the true altitude of a meteor. The path of the meteor is given by the line AB. Observer A sees it apparently moving from a′ to b′; observer 2, from a² to b². This gives all the information needed for working out the altitude of the meteor above the ground.

that it has not affected the value of amateur work, though the naked-eye watcher can still make himself very useful indeed.

Sporadic meteors are frequent enough, and the watchful observer will seldom fail to record fewer than five or six per hour, but it is far more entertaining (though not necessarily more useful) to concentrate upon definite showers. Occasionally there will be so many meteors in quick succession that the observer will be hard pressed to record them all, but this will not happen except near the maximum of a rich shower, and there are bound to be long periods of patient waiting.

Note that visual meteors are twice as plentiful in the period from midnight to 6 a.m. than from 6 p.m. to midnight. In the evening, we are on the 'rear side' of the Earth as it moves along in its orbit, so that meteors have to catch us up; in the morning hours we are on the 'front side', so that meteors meet us coming. Not only are more meteors to be expected after midnight than before it, but also the morning meteors will have greater relative speed, just as a car moving at 20 mph and meeting a second car moving at 40 mph will be badly damaged if the collision is head-on, but only bumped if rammed from behind. It is the relative speed of a meteor which is the main factor in its brightness, so that the morning meteors will be more brilliant and thus easier to record.

Meteor photography is a most interesting pursuit, and it can be carried out with an

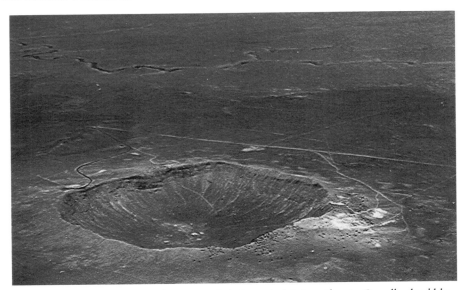

Fig. 11.8. The Meteor Crater in Arizona, as I photographed it from the air. (It really should be called the Meteorite Crater.)

ordinary camera. I have given some details of procedure in the Appendix 29. In particular, meteor spectra are most important, and some of the best results to date have been obtained by amateurs.

Larger objects may encounter the Earth, and if they complete their journey without being burned away they are known as meteorites. There is a popular misconception here. A meteorite is not simply a large meteor – in fact there is no connection at all. Meteors come from comets; meteorites seem to originate in the asteroid zone, and, as I have already said, there is probably no difference between a large meteorite and a small asteroid. Most museums have meteorite collections, and an expert can soon tell whether a piece of material is meteoritic or not, though the layman can easily be misled. Basically, meteorites are of two main types, stones (aerolites) and irons (siderites), though there are also many intermediate varieties.

Large meteorites are rare, though now and then they can produce craters. The most famous of these craters is in Arizona, not too far from Flagstaff; it is almost a mile wide, and was produced by an impact which took place well over 20000 years ago (Fig. 11.8). It is a well-known tourist centre, and I advise you to go and see it if you have the chance; there is no difficulty in getting there, since it can be approached by a road leading off Highway 99.

One very interesting fall occurred in England on Christmas Eve, 1965. A meteorite flashed across the Midlands, attracting considerable attention, and broke up; fragments of it came down near the village of Barwell, in Leicestershire (I even found one myself, when I visited the site some time later). The original weight must have been around 200 pounds, which is a British record. One fragment went through the window of a house in Barwell, and was subsequently found nestling comfortably in a vase of artificial flowers.

Since then we have had the Bovedy Meteorite of April 25 1969, which shot across England and Wales and dropped fragments in Northern Ireland, though the main mass apparently fell in the sea. I have to admit that I missed it by about two minutes. I had been in my observatory at Selsey in Sussex, observing variable stars, and had just gone indoors to change my charts when the meteorite passed over. So far as I am concerned, this was yet another instance of Spode's Law. Mind you, meteors of this kind are puny when compared with some of the giants. The holder of the heavyweight record is the Hoba West Meteorite in Southern Africa; it fell in prehistoric times, and has an estimated weight of at least 60 tons.

So much, then, for the Solar System. Now let us turn our attention toward the stars.

12

The Stellar Sky

Look up into a dark, clear sky and you will see hundreds of stars. It was natural for them to be grouped into patterns or constellations, and they were given attractive names, though different civilizations used different systems — thus the Chinese constellations were quite different from those of the Egyptians or the Greeks. We use the Greek system, suitably modified. There are gods, demigods, heroes and even everyday objects; Orion, the Hunter, is very much in evidence, as are Hercules, Perseus and many others, as well as the much more down-to-earth Triangle and Cup.

The names are generally used in their Latin forms, and I propose to do so in this book; there is nothing abstruse about them. A full list, with the English equivalents, is given in Appendix 16.

Ptolemy of Alexandria, last of the great astronomers of ancient times, listed 48 constellations; others have been added since, and today the grand total of accepted constellations is 88, though in size and importance they are very unequal. It must be admitted that some of them seem unworthy of separate existence, and few bear any resemblance to the objects after which they are named. (It takes a very lively imagination to make a crab out of Cancer, or a ram out of Aries!) Further proposed additions with barbarous names such as Sceptrum Brandenburgicum, Officina Typographica and Lochium Funis have been mercifully forgotten, and one of Ptolemy's originals, Argo Navis (the Ship Argo) was so huge and unwieldy that it has been split into three separate parts.

Probably the most famous of all the constellations are Ursa Major (the Great Bear), Orion, and Crux Australis (the Southern Cross). Of these, Ursa Major — containing the Plough or Dipper pattern — lies in the far north of the sky, so that over Britain it never sets, while Orion is crossed by the celestial equator, and Crux is so far south that from Europe it can never be seen at all, though it is familiar tb almost all Australians, New Zealanders and South Africans. Stars which never set are said to be 'circumpolar', so that Ursa Major is circumpolar over Britain (Fig. 12.1).

Before going any further, I must say something about right ascension and declination. Broadly speaking, these are the celestial equivalents of longitude and latitude on the Earth's surface, though there are differences in detail.

Declination is reckoned in degrees north or south of the celestial equator, while the

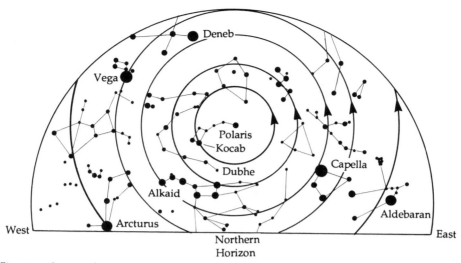

Fig. 12.1. Circumpolar and non-circumpolar stars. The star near the pole is, of course, the Pole Star. Stars near it, such as Ursa Major, never set as seen from Britain; stars further away, such as Arcturus, rise and set.

equator itself is merely the projection of the Earth's equator on to the celestial sphere (Fig. 12.2). The north celestial pole has a declination of 90 degrees north (+90°); Polaris, the Pole Star, has declination of over +89 degrees, so close to the actual pole that it remains almost motionless in the sky. Observers in the southern hemisphere are not so lucky, since there is no bright star anywhere near the south celestial pole.

To anyone living at the North Pole, Polaris would appear to stay almost overhead. From Greenwich (latitude $51\frac{1}{2}°$ N) the altitude of Polaris is also $51\frac{1}{2}°$; from the equator (latitude 0°) Polaris has no altitude at all – in other words, it lies right on the horizon. South of the equator, Polaris can never be seen, because it never rises above the horizon (Fig. 12.3).

If you want to work out whether some particular star is circumpolar, or whether it is permanently invisible, all you have to do is to subtract your latitude from 90 degrees and then compare it with the declination of the star. For example, consider my own observatory in Selsey, Sussex, where the latitude is 51°North. 90 − 51 = 39. Therefore, any star north of declination +39° will never set, and any star south of declination −39° will never rise. I can never see the bright southern star Canopus, whose declination is −53°, but I can always see Alkaid in Ursa Major, where the declination is 49°. From Invercargill in New Zealand (latitude 46° S) Canopus is always on view, Alkaid never. In Appendix 23 I have given the declinations of some of the brightest stars.

The point at which the Sun crosses the celestial equator, moving from south to north, is known as the Vernal Equinox, or First Point of Aries. The Sun reaches it around March 21 each year, when its declination is, naturally, zero. Six months later it crosses the equator again, this time moving from north to south, at the Autumnal Equinox or First Point of Libra. The Vernal Equinox is to the sky what the Prime

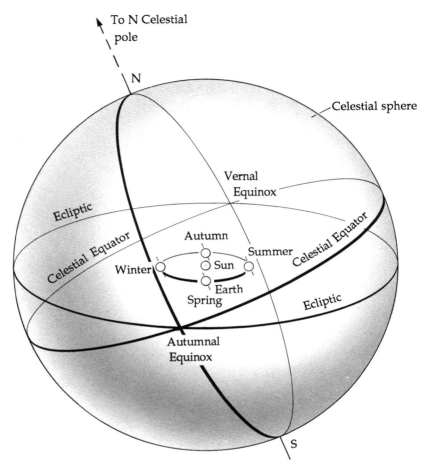

Fig. 12.2. The Celestial Sphere.

Meridian is to the Earth – but it is a point which has been chosen for us by Nature, whereas Greenwich was selected as the zero for longitude merely because the famous Observatory happened to have been built there.*

The angular distance of a star eastward of the Vernal Equinox is known as the star's right ascension. It can be given in degrees, but is more usually measured in hours, minutes and seconds of time.

To make this clear, I must refer you to the 'meridian' of any observing point on the Earth; it is the great circle on the celestial sphere which passes through the celestial pole and also the zenith, or overhead point. Obviously, a star on the meridian will be at its maximum height above the horizon, and is said to 'culminate' (Fig. 12.4). The First Point of Aries must pass across the meridian at any place once in every 24 hours (sidereal time), and the difference between this time and the time of the star's meridian

---

* Greenwich was accepted as the zero meridian by international agreement almost a hundred years ago. The only countries to object were (naturally) France and Ireland.

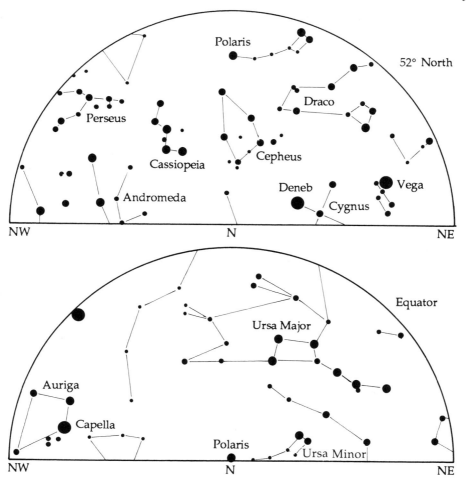

Fig. 12.3. (Upper) From London (latitude approximately 52 degrees) Polaris is 52 degrees above the horizon. From the equator (latitude 0) Polaris has an altitude of 0 degrees – in other words, it lies on the horizon. (For this illustration we may safely ignore the fact that Polaris is not exactly at the north celestial pole.)

passage will give the right ascension of the star. For instance, the brilliant Sirius reaches the meridian 6 hours 45 minutes 08.9 seconds after the First Point has done so, indicating that the right ascension of Sirius is 6 h 45 m 08.9 s.

The slight shift of the celestial pole due to precession, described in Chapter 2, means that a star's right ascension and declination alter very gradually over the years. In this book the positions are given for the year 2000, and it will be a long time before the error becomes great enough to be worrying. Incidentally, the First Point of Aries has now moved into the neighbouring constellation of Pisces, the Fishes, though we still keep to the old name.

A telescope equipped with setting circles and clock drive can be swung to any

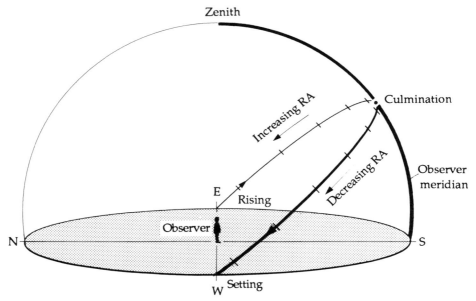

Fig. 12.4. Culmination of a star.

desired right ascension and declination, so that as soon as the position of the target object is known it can be found at once. It is obvious that while the right ascensions and declinations of stars do not alter much except over long periods, those of the Sun, Moon and planets will change quickly.

Dividing the stars into constellations, and naming the brightest of them, is enough for a rough classification. Most of the brilliant stars have proper names, such as Sirius, Canopus, Rigel, Vega and Capella. However, it would be a hopeless task to give special names to each star, and we have recourse to letters or numbers.

A system introduced in 1603 by the German astronomer Johann Bayer has stood the test of time so well that it will certainly never be altered. On this system, the leading stars of each constellation are given Greek letters, beginning with Alpha for the brightest star and working through to Omega, the last letter of the Greek alphabet. Thus in Aries, the Ram, the brightest star is Alpha Arietis (Alpha of the Ram), the second brightest is Beta Arietis, the third brightest Gamma Arietis, and so on. Unfortunately the strict order is often not followed, and in some constellations the order is chaotic. In Orion Beta is the brightest star, followed by Alpha, Gamma, Epsilon, Zeta and then Kappa, with Delta an 'also ran'; in Sagittarius, the Archer, the two leading stars are Epsilon and Sigma, while both Alpha and Beta Sagittarii are comparatively faint. A list of the Greek letters, with their English names is given in Appendix 20.

This system is all very well, but it can cope with only the 24 principal stars in each constellation, which in some cases (such as Orion) is not nearly enough. John Flamsteed, the first Astronomer Royal, preferred to give the stars numbers, beginning in each constellation with the star of lowest right ascension. Still fainter stars, not listed by Flamsteed, have been given numbers in later catalogues, and the result is that all the bright stars have several designations; for example, Rigel in Orion is known also as Beta Orionis and as 19 Orionis. In general, proper names are used for only the twenty or thirty brightest stars, plus a few others of particular interest.

Next, we need to divide the stars into grades or magnitudes of apparent brilliancy. The scale works rather in the manner of a golfer's handicap; the brighter the star, the lower the magnitude – 1 brighter than 2, 2 brighter than 3, and so on. The faintest stars normally visible with the naked eye on a clear night are of magnitude 6, while the world's most powerful telescopes can go down to at least 25. Modern measuring instruments known as photometers can give the magnitudes very accurately; thus Alkaid in Ursa Major is of magnitude 1.86.

A few stars are actually brighter than magnitude 1.0, so that they have values of less than unity; examples are Rigel (0.08) and Altair (0.77), so that Rigel is appreciably the brighter of the two. Four stars – Sirius, Canopus, Alpha Centauri and Arcturus – have minus magnitudes; Sirius is −1.46. On this scale the planet Venus can reach magnitude −4.4, while the Sun is −26.8. The scale is not so arbitrary as might be thought; a star of magnitude 1 is exactly a hundred times as bright as a star of magnitude 6.

The stars are of different luminosities, and are at different distances from us, so that our constellation patterns are due to nothing more significant than line of sight effects. In Ursa Major, for instance, one of the seven bright stars (Alkaid) is more remote than the other six, while Polaris, in Ursa Minor, is six hundred times as far away from us as Alkaid – so that Polaris is much more remote from Alkaid than we are. Do not imagine that any two stars have any real connection with each other merely because they happen to lie in the same constellation!

For many years astronomers were quite unable to measure the distances of the stars. It was a problem which defeated even the great William Herschel. It was finally solved, sixteen years after Herschel's death, by the method of parallax.

The best way to demonstrate parallax is to make a practical experiment with a pencil, holding it at arm's-length and looking at it with alternate eyes. First, line up the pencil with an object such as a clock on the mantelpiece, or a tree in the garden. Now, without moving your face or the pencil, shut your first eye and open the other. The pencil will no longer be aligned with the clock or the tree; it will seem to have shifted – because you are looking at it from a slightly different direction; your two eyes are not in the same place. If you know the distance between your eyes, and also the angular shift of the pencil, you can work out the distance of the pencil from your face by using fairly straightforward mathematics. The apparent shift of the pencil is a measure of the parallax.

Exactly the same principle can be used to measure the distance of a comparatively nearby star against the background of more remote stars, but the base-line has to be enormously long. Fortunately Nature has given us such a base-line; the Earth swings from one side of the Sun to the other in a period of six months, shifting in position by 186 million miles (Fig. 12.5). If S is our 'near' star, it will appear at position S1 in January, but S2 in July, so that if we measure the angular shift we can find the distance.

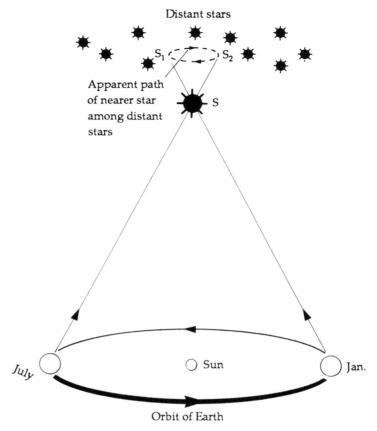

Distant stars

Apparent path
of nearer star
among distant
stars

Fig. 12.5. The parallax method of measuring the distance of a star. In January, the relatively close star will be seen at S¹; in July, it will be seen at S²; the amount of apparent displacement (S1–S2) gives the parallax and, hence, the distance of the star S.

The parallax shift is so tiny that it is hard to measure, and there are numerous corrections to be made, but there is nothing complicated in the method itself, and in this way Friedrich Bessel, in 1838, managed to measure the distance of the fifth-magnitude star 61 Cygni. The parallax is 0.29 of a second of arc, corresponding to a real distance of 11.2 light-years or roughly 66 million million miles. Only three naked-eye stars are closer than that.

The parallax method breaks down for more remote stars, because the shifts become too small to be properly measured. At 160 light-years the method has become untrustworthy, and at 600 light-years it is quite useless. Indirect methods have had to be developed, and most of these involve finding out the actual luminosity of a star as compared with the Sun, since as soon as know both the real brilliance and the apparent magnitude we can find the distance – much as we can judge the distance of a lighthouse out to sea if we know the power of the lamp.

Even the nearest of all the stars beyond the Solar System, Proxima Centauri in the

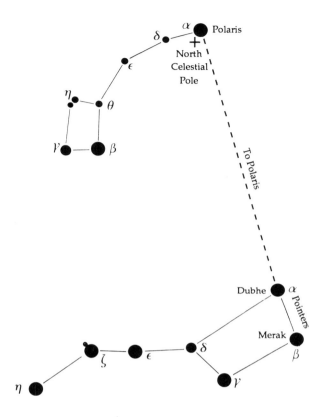

Fig. 12.6. The Pointers in Ursa Major (Merak and Dubhe) show the way to the Pole Star in Ursa Minor, which is within one degree of the north celestial pole.

southern sky, is immensely remote. It lies at 4.2 light-years, or over 24 million million miles. Most of the naked-eye stars are much further away; Sirius has a distance of 8.6 light-years, Rigel in Orion, over 900 light-years, and Deneb in Cygnus (the Swan) as much as 1800 light-years. Look at Deneb now, and you see it as it used to be at the time when the Romans still occupied Britain. Once we go beyond the Solar System, our view of the universe is bound to be very out of date.

The stars have a wide range of luminosities; we know of celestial searchlights more than a million times as powerful as the Sun, while others are mere glow-worms with only a tiny fraction of the Sun's luminosity. They also differ in colour; some are bluish, some white, others yellow or red. The colours indicate different surface temperatures. For example, Rigel in Orion is white, while Betelgeuse is orange-red, so that Rigel is

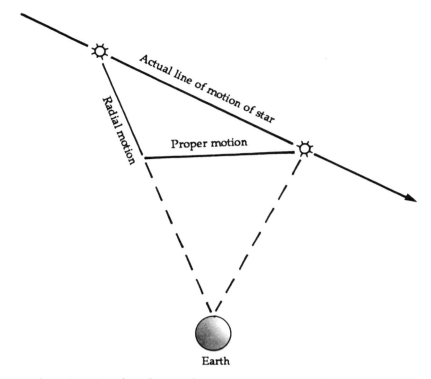

Fig. 12.7. *A star is moving through space; the motion as seen from Earth is a combination of the transverse or* proper *motion, and the toward-or-away or* radial-motion.

the hotter of the two; but to make up for this, Betelgeuse is much the larger. Its huge globe could swallow up the whole orbit of the Earth round the sun.

To the naked-eye observer, the stars seem to stay in the same relative positions. Two of the stars in Ursa Major, Dubhe and Merak, point to the Pole Star (Fig. 12.6); they have done so for generations, and will continue to do so for generations more. Of course, the old term 'fixed stars' is misleading. The stars are moving around at high speeds, but they are so remote that it takes centuries for them to show any obvious shifts in position. Yet over the ages, the shifts will mount up, and eventually the two Pointers will no longer show the way to Polaris.

The slow movement of a star against its background is known as the star's proper motion; do not confuse it with the slight yearly shifts due to parallax. The holder of the speed record is Barnard's Star, a faint red sun at a distance of 6 light-years. Even so, it takes Barnard's Star 190 years to shift across the sky by a distance equal to the apparent diameter of the full moon.

There is also a movement in the line of sight, termed the radial motion (Fig. 12.7). If a star is coming straight toward us or straight away from us, it will show no proper motion at all, but its radial motion will be detectable by means of the spectroscope.

The Sun is an ordinary star, and other stars show spectra of much the same type.

Temperature differences and other factors will cause modifications, but usually there will be a continuous rainbow crossed by dark absorption lines. If the star is approaching us, the dark lines will be shifted slightly toward the short-wave or violet end of the spectrum, while if the star is receding the shift will be toward the red. By measuring the shifts of the lines, we can work out the radial velocity of the star.

There is a good way to show what is meant. Listen to an approaching ambulance or police car, moving quickly with its siren sounding. You will note that the sound is high-pitched when the vehicle is approaching; more sound-waves are reaching your ear than would be the case if the vehicle were standing still, so that the wavelength is effectively shortened. When the vehicle has passed by, and has started to recede, fewer sound-waves per second will reach you, and the note of the siren will drop. Light is a wave motion, and when the source is moving away the 'pitch' is shifted toward the long-wave end of the spectrum. This is the well-known Doppler Effect, so called because it was first described by the Austrian physicist of that name more than a hundred years ago.

Sweeping the night sky with a telescope, or even binoculars, is a fascinating pastime. Some of the stars show vivid colours; some are double, others multiple. There seems to be no end to it all, and no observer can hope to cover the whole of the sky in the course of a lifetime. The more you see, the more you will realize that our Solar System is a tiny unit in space.

13

The Nature of a Star

Look at a bright star through a telescope, for the first time, and you may expect to see a huge globe filling most of the field of view. I can only say that if you do in fact see anything of this kind, there is something wrong with your telescope. Not even the world's greatest instruments can show a star as anything but a point of light. This is not because they are small — far from it; as we have seen, some of them are big enough to contain the whole path of the Earth round the Sun. Distance is the trouble. If we want to find out much about the nature of the stars, we must use less direct methods, based generally upon the principle of the spectroscope.

I have already said something about the spectrum of the Sun; and since the Sun is an ordinary star, it is only to be expected that other stars show spectra of the same basic type. Yet there are tremendous differences in detail, because the stars show a wide range of luminosity and mass as well as size. The colour of a star is a key to its surface temperature, with the steely blue of Vega contrasting with the pure white of Sirius, the yellow hue of Capella or the orange-red of Betelgeuse. Over a century ago Angelo Secchi, one of the great pioneers, found that there were well-marked spectral classes; for example, white stars of the Sirius type showed prominent dark absorption lines due to hydrogen, while with Rigel the helium lines were dominant, and the spectra of orange and red stars showed bands due to molecules. Secchi divided the stars into four classes. A more elaborate system, devised by E.C. Pickering at Harvard in the United States, increased the number to eleven with one type merging gradually into the next.

On the Harvard system, each type of star is given a letter of the alphabet. Originally it was intended to take the usual sequence of letters, but some of the early classes were found to be either unnecessary or else out of order, and the final sequence was alphabetically chaotic: W, O, B, A, F, G, K, M, R, N, S. (I think most people know the mnenonic 'Wow! O Be A Fine Girl Kiss Me Right Now Sweetie'.) The sequence indicates decreasing surface temperature, W stars being the hottest and R, N and S stars the coolest. The Sun, as befits its undistinguished character, comes in Type G, near the middle of the series. Each type is divided into sub-grades from nought to 10, so that, for example, a star of type A5 is midway between A0 and F0, while there is not much difference between F9 and G0.

I repeat that this is not a book about theoretical astronomy, but I must say a little about the characteristics of the various spectral types, because the observer will need to know them. First in the order are the W or Wolf–Rayet stars, with surface temperatures of up to 80 000°C. They have bright lines in the spectra as well as dark ones; all are highly luminous, and very remote; all are unstable, with expanding shells of material moving out from their surfaces. They are rare, and less than 200 have been found in our Galaxy. O stars, with surface temperatures from 35 000–40 000 degrees,* are also rare and luminous.

B-type stars, such as Rigel, are bluish-white, with dominant lines of hydrogen and (particularly) helium. Stars of type A are pure white, with prominent hydrogen lines and surface temperatures of from 8000 to 10 000°C. F-type stars, with conspicuous calcium lines, are slightly yellowish, and stars of type G are obviously yellow; their lower surface temperatures, of no more than 6000°C, mean that metallic lines start to become evident. These are even more pronounced with the orange K-stars, such as Arcturus in Boötes. Stars of type M have very complicated spectra, with many bands due to molecules –not seen with the hot stars, as molecules would be broken up by the high temperatures. Finally there are the coolest stars, with surface temperatures of from 2500 to 2600 degrees, making up the comparatively rare types R, N and S. Most of these are variable in brightness.

It may be convenient to group the stars in this way, but there are some complications which seem rather unexpected at first sight. Consider, for instance, two M-type stars, the brilliant Betelgeuse and the dim Wolf 359. Betelgeuse shines as brightly as 15 000 Suns put together, while Wolf 359 has only 1/50 000 of the Sun's power, so why should we class them together? To say the least of it, they are ill-assorted companions.

In the early years of the twentieth century, two astronomers, Ejnar Hertzsprung of Denmark and Henry Norris Russell of the United States, drew up diagrams in which the stars were plotted according to their spectral types and their luminosities (Fig. 13.1). On these diagrams – always known today as H–R Diagrams – it was obvious at once that apart from types W, O, B and A, the classes were divided into 'giants' and 'dwarfs'. We can find M-type giants such as Betelgeuse, and M-type dwarfs such as Wolf 359, but M-stars about equal in power to the Sun do not seem to exist at all. When it became possible to estimate the diameters of the stars, the effect was even more pronounced. If we picture a scale model and make Betelgeuse a globe with a diameter equal to that of a cricket pitch, Wolf 359 will be no larger than a croquet ball. It is true that the range in mass is not so great, because the dwarfs are denser than the giants, but the distinction between giants and dwarfs is still quite unmistakable.

The recognition of the giant and dwarf classes was followed by a very simple, straightforward theory about the life-history of a star. It was assumed that when it first condensed out of the interstellar gas and dust, a star was hardly hot enough to emit visible light. Naturally it would start to contract, because gravity would tend to pull all its mass together; this would cause heat, so that the star would become a red giant such as Betelgeuse. As the contraction went on, the star would become an orange

---

* From here onward I have given all temperatures in degrees Centigrade (or, if you like, Celsius). If you want to use Fahrenheit, I have given a conversion table in Appendix 21.

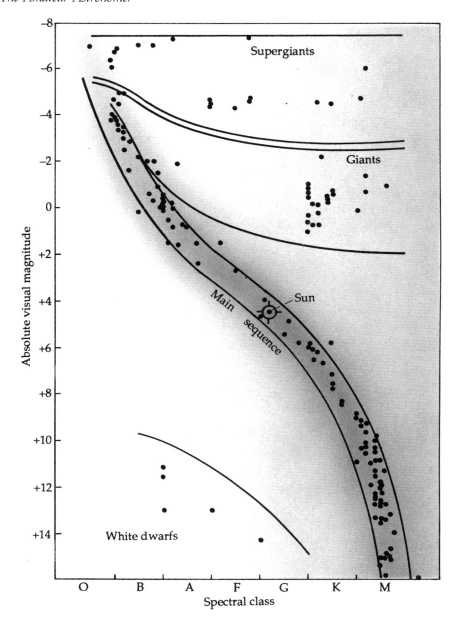

Fig. 13.1. The Hertzsprung–Russell or H–R Diagram. The Main Sequence runs from top left to bottom right; the giants and supergiants are to the upper right, and white dwarfs to the lower left.

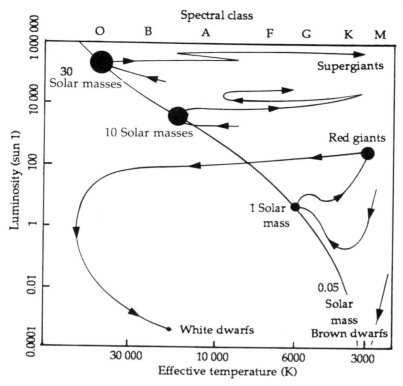

Fig. 13.2. Evolutionary tracks on the H–R Diagram for stars of various masses.

giant (type K) and then a yellow giant (type F) before passing through type G and joining the top left of the H–R diagram at type W, O or B (Fig. 13.2). This would be the peak of a star's career. It would go on shrinking, but it would also start to cool, since its main energy would have been spent. It would pass down the dwarf series or 'Main Sequence', becoming first an F-dwarf, then a G-dwarf such as the Sun, and then a small, red star of one of the later types, finally losing all its heat and changing from a dim red dwarf such as Wolf 359 into a cold, dead globe – a black dwarf.

It all sounded beautifully simple – a progression from 'early' type stars (W, O, B) through to 'late' type (M and the less common R, N and S). The energy of a star was originally believed to be due to simple gravitational contraction. Unfortunately there were all manner of objections, and by now we know that the life-story of a star is very different. What happens to it depends mainly upon its initial mass. We have already noted that the Sun's energy is produced by nuclear transformations going on deep inside it – mainly the conversion of hydrogen into helium, admittedly by a somewhat roundabout method which need not concern us at the moment. Stars of different mass evolve differently, and end their careers in very different ways.

With a star of very low mass, the core temperature never reaches a value high enough for nuclear reactions to begin at all. Therefore, the star simply glows feebly for an immensely long period before fading away to become a cold, dead black dwarf (though whether the universe is old enough for any black dwarfs to have been formed as yet is by no means certain). But with the Sun, the globe will have to re-arrange itself drastically as soon as the available hydrogen has been used up, and much of it has been turned into helium. The outer layers expand and cool, while the core shrinks and heats up; the star becomes a red giant. Eventually it throws off its outer layers completely, going through the stage which we call a planetary nebula. What is left of the star itself is left exposed as a small, amazingly dense object of the kind known as a white dwarf.

White dwarfs have been described as bankrupt stars, shining only because they are still slowly contracting. They are certainly plentiful, but they are so faint that they are not easy to detect unless they are relatively close to us. Their surfaces are not cool, and have temperatures sometimes above 10000 degrees, so that their dimness proves that they must be very small. For example, the white dwarf companion of the brilliant Sirius has a mass equal to that of the Sun, but a diameter of no more than about 26000 miles, smaller than that of a planet such as Uranus or Nepture. The atoms in it are broken up and crushed together, with almost no waste space, so that the density must be of the order of 60000 times that of water. Several tons of white dwarf material could be backed into a thimble.

Our Sun will become a white dwarf one day, but at least we have the sombre satisfaction of knowing that we will not be there to see it die. The Earth cannot possibly survive the blast of radiation which the Sun will emit during its red giant stage. If Mankind still exists at that remote epoch, it will be time for us to make a move!

With a star of more than 1.4 times the initial mass of the Sun, everything happens at an accelerated pace. Things proceed much as before, though more quickly, until the end of the helium conversion, but then different reactions take over; the central temperature soars to a fantastic value, and heavier and heavier elements are built up. Eventually, the core consists largely of iron. This is the beginning of the end, because iron will not react. Energy production stops abruptly; there is an 'implosion' (the opposite of an explosion) followed by a shock-wave, and the star blows itself to pieces in what we call a supernova outburst. I will have more to say about supernovæ in Chapter 15.

What if the star's mass is greater still – more than about eight times that of the Sun? The star cannot even 'go supernova'. When the collapse starts, it is so sudden and so catastrophic that nothing can halt it. As the star becomes smaller and smaller, denser and denser, the escape velocity goes up, until it reaches a value of 186000 miles per second. This, of course, is the velocity of light, so that not even light can break free from the old star – and if light cannot do so, then certainly nothing else can, because light is the fastest thing in the universe. The collapsed star is left surrounded by a region of space from which nothing – absolutely nothing – can escape. It has produced a black hole.

Obviously we cannot see black holes, and we can detect them (we think!) only because of their effects upon visible objects. To go into more detail would be beyond the scope of a book devoted to amateur work. So let me turn now to things which we can actually see – and there is plenty of variety in the starlit sky.

14

Double Stars

Of all the constellations in the northern sky, undoubtedly the best-known is Ursa Major, the Great Bear, with its characteristic Plough or Dipper pattern. It is not so brilliant as Orion or so spectacular as the Southern Cross, but it is quite unmistakable, and from countries such as Britain it is always on view whenever the sky is dark and clear.

Even a casual glance will show that there is something unsual about the 'second star in the tail', Mizar or Zeta Ursæ Majoris. Mizar itself is of the second magnitude; close beside it is Alcor, which is only of the fifth magnitude and is none too easy to see under conditions of mist or moonlight.

Double stars are very common in the sky, though most of them have components so close together that they cannot be separated with the naked eye. Telescopically they are often spectacular, with components of different colours – and they are also useful for testing the light-grasp and quality of a telescope.

There are two classes of double stars. First there are the so-called optical pairs, whose components are not really associated; one happens to lie almost behind the other, so that we are dealing with nothing more than a line-of-sight effect. Yet rather surprisingly, optical pairs are very much in the minority. Most doubles are physically-associated or 'binary' systems.

The existence of binary pairs was recognized almost two hundred years ago by Sir William Herschel, more or less by chance. He had been trying to measure star-distances by the method of parallax, and he had made observations of double stars because he expected – naturally – that the closer member would show a parallax shift compared with the further. He never managed to measure the distance of any star, because his telescopes were not accurate enough, but he did find that in some cases the components of a pair were in orbital motion round each other. Today we know that binaries may even be more common than single stars.

It is misleading to say simply that the less massive member of a binary pair revolves round its companion. Though the two may be very unequal in size and brilliance, they will not be so unequal in mass; as we have noted, the mass-range is comparatively slight, and with some pairs the smaller component may well be the more massive of the two. What happens is that the two bodies move round their common centre of

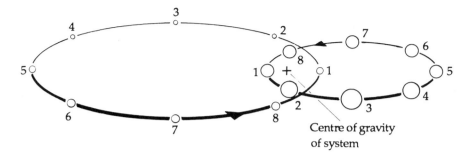

Fig. 14.1. *Movements of the components of a binary star system. The centre of gravity of the system will be displaced toward the more massive component.*

gravity. If they are equal in mass, the centre of gravity will be half-way between them; if one star is more massive than the other, the centre of gravity of the system will be displaced toward it (Fig. 14.1). The situation is basically the same as with the Earth and the Moon, though the centre of gravity of the Earth–Moon system actually lies inside the Earth's globe.

A pair of binoculars will show many pairs, and a small telescope will reveal dozens. Sometimes the components are equal, so that they look like identical twins; such as Alya or Theta Serpentis, in the Serpent, and Mesartim or Gamma Arietis, in the Ram. In other cases one star is much brighter than the other, and may drown its companion to such an extent that a telescope of considerable power is needed to show both independently. Sirius is a good case of what I mean. The white dwarf companion has only 1/10000 the brightness of Sirius itself. Its apparent magnitude is 8.5, so that if it could be seen shining on its own it would be within the range of good binoculars, but I admit that I have never been able to make it out with any telescope of less than 10 inches aperture, though observers with keener eyes than mine will undoubtedly do better.

Binary stars are of great importance. Their orbits can be worked out; and as soon as the distance between the components is known, together with the period of revolution, the combined mass of the system can be found. Suppose, for instance, that we find a pair in which the mean distance between the components is 93 million miles, and the revolution period is one year. The Earth moves round the Sun at this distance and in this period, so that the combined mass of the Earth–Sun pair would be equal to the combined mass of the two stars in the binary. In practice we can forget about the Earth, whose mass is negligible by stellar standards, so that the two members of the binary system would, together, equal the mass of the Sun.

The whole method depends upon careful measurements of the relative movements of the components, and it is not surprising that astronomers have paid great attention to them; it takes years for the shifts to become noticeable – at least in most cases. Two measurements are needed: the separation, in seconds of arc, and the position angle. The position angle of a double star (binary or otherwise) is the direction of the fainter member (B) as reckoned from the brighter (A), beginning at 0 degrees at the north point and going round by east (90°), south (180°) and west (270°) back to north (360°

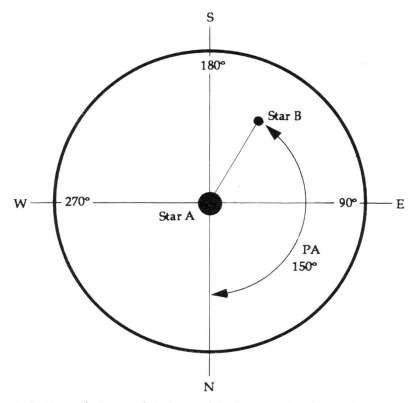

Fig. 14.2. Position angle, from north (0 degrees) through east, south and west. The position angle of the fainter component is measured with reference to the brighter.

or 0°), as shown in Fig. 14.2. Published position angles and separations are generally good enough to allow the observer to form a mental picture of the pair before actually going to the telescope, though in the case of 'twins' it is not easy to decide which of the components is meant to be the senior partner.

Measuring the separations and position angles of double stars is something which cannot really be attempted with a telescope of below 6 inches aperture; you must also have an equatorial mounting, a driving clock and an instrument known as a micrometer. Not many amateurs tackle this sort of work, but there is plenty of scope for anyone who is suitably equipped. It is also true that even with bright pairs, some of the values given in published lists are out of date, and could well do with revision.

Mizar, of course, is the 'classic' double. The separation between Mizar itself and Alcor is about 700 seconds of arc, but even a small telescope will show that the bright star is again double, with rather unequal components separated by 14.5 seconds of arc (Fig. 14.3). They make up a genuine binary, but are so far apart in space that their relative shifts are very slow. Between Alcor and the bright pair – or, more accurately, forming a triangle with them – is a much fainter star, of the 8th magnitude. It has been nicknamed Sidus Ludovicianum, because in 1723 the courtiers of the Emperor Ludwig

Fig. 14.3. Mizar, Alcor, and Ludwig's Star, drawn by Paul Doherty.

V apparently believed that it had appeared suddenly! This, of course, is out of the question; it is a perfectly normal star, but is not a true member of the Mizar system, and merely lies in the background. I can just see it with powerful binoculars, and any small telescope will show it easily.

Actually, there is a minor mystery here. The old Arab astronomers called Alcor 'a test for keen eyes', but today there is nothing difficult about it when the sky is clear, and in no sense can it be regarded as a test. Either there has been some mistake in translation, or Alcor has brightened up during the past thousand years – or else the Arabs were not referring to Alcor at all. It has been suggested that their 'test' was Ludwig's Star, which in that case must be variable. I confess that I am sceptical; all the same, it is decidedly curious.

There are various other naked-eye doubles – for instance, Theta Tauri in the cluster of the Hyades – and if you want a pair of 'identical twins', separable with a very small telescope, I recommend Alya or Theta Serpentis, not far from the prominent constellation of Aquila, the Eagle. Both are of the fourth magnitude, and are just over 22 seconds of arc apart. It is said that one component is slightly brighter than the other, but I have never been able to see any difference between them; both are white, of spectral type A5. They are physically associated, but are so far apart that their revolution period must be many centuries.

A fainter pair is the famous 61 Cygni, in the Swan, which was the first star to have its distance measured (by Friedrich Bessel, in 1838). The magnitudes are 5.3 and 5.9, so that the pair is by no means prominent; the separation is 30 seconds of arc, and both components are red dwarfs of type K. Even when put together, they barely equal the luminosity of the Sun. 61 Cygni is a binary, but the period has been estimated as over 650 years.

Arich or Gamma Virginis, not far from the celestial equator, is of rather different type. Again we have almost equal components, both of magnitude 3.5, but the revolution period is only 179 years, so that changes in separation and position angle are noticeable over the course of a lifetime. When I first looked at Arich, around 1930, it was one of the easiest telescopic doubles in the sky, with a separation of some 6 seconds of arc. The components are now much closer together – not because they are really so, but because we are seeing the pair from a narrower angle. The present separation is only 3 seconds of arc. By the year 2010 it will be down to less than 1 second of arc, so that a telescope of some power will be needed to show the stars separately. On the other hand Castor, in Gemini (the Twins) is becoming easier; the separation was only just over 2 seconds of arc in 1960, but has now grown to 3, and will reach 5.6 seconds of arc by the year 2020, so that Castor will then be separable with any telescope. The magnitudes are 1.9 and 2.9, and the period is 511 years.

Two of the most brilliant stars of the far south are fine doubles. Alpha Centauri is the third brightest star in the whole of the sky; it is surpassed only by Sirius and Canopus. The components are of magnitudes 0.0 and 1.4, so that with the naked eye Alpha Centauri has a magnitude of − 0.3. Since the revolution period is only 80 years, both separation and position alter quite quickly. The separation can range between 2 seconds of arc and 22 seconds; at present it is almost 20 seconds, though it is decreasing. The brighter component is of type G, like the Sun, while the fainter though larger member of the pair is orange, and of type K. Alpha Centauri is only 4.3 light-years away, and the faint third member of the system, Proxima, has the distinction of being the nearest of all stars beyond the Sun. Do not expect to see Proxima with a small telescope; it is not much above the eleventh magnitude, and is none too easy to identify even when you know just where it is.

The other magnificent southern double is Acrux or Alpha Crucis, leader of the Southern Cross. The magnitudes are 1.4 and 1.9; both components are white, or type B, and the separation is over 5 seconds of arc. Acrux is a physically-associated pair, but the revolution period is very long. Also in the same low-power field is a third star, of magnitude 5.

For a very different kind of double, look at Antares, in Scorpius (the Scorpion). Here

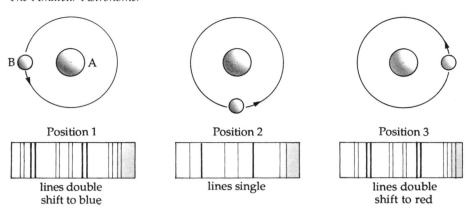

Fig. 14.4. Principle of a spectroscopic binary. In positions 1 and 3 the lines are doubled; in position 2 they are single.

we have a vast red supergiant of type M, so big that it could more than hold the Earth's path round the Sun; it has a companion of magnitude 6.5 at a distance of 3 seconds of arc. This sounds as though it should be easy to split, but in fact it does need a telescope of some size, because the companion is so overpowered by the brilliance of Antares itself. I can see both with a 6-inch reflector, but I find them very difficult with any smaller aperture. The companion is of type B, but tends to look greenish by contrast with the red supergiant. The revolution period is around 800 years, and the real separation between the two is at least 500 times as great as the distance between the Earth and the Sun.

The most beautiful of all double stars are those which show contrasting colours. Pride of place must surely go to Albireo or Beta Cygni, the faintest of the five stars making up the 'cross' of the Swan. The main star is of the third magnitude, and is golden-yellow, while the fifth-magnitude companion appears vivid blue. The separation is 34 seconds of arc, so that both stars can just about be seen with powerful binoculars; any small telescope will show them well, and in my view, at least, Albireo is one of the 'show-pieces' of the sky.

If you want to try your luck with a really difficult double, try Delta Equulei in the little constellation of the Foal, not far from Cygnus. Both components are F-type dwarfs; the period is 5.7 years, and the separation ranges between 0.1 and 0.3 seconds of arc. Delta Equulei is 49 light-years away.

Some double stars are too close to be split with any telescope, but can be detected spectroscopically by means of our old and reliable ally, the Doppler effect. In the very over-simplified diagram in Fig. 14.4, it is assumed that the fainter star (B) is moving round the brighter (A). In position 1, B is moving toward us, and the lines in its spectrum will show a violet shift; in position 2, B is crossing the field and so exhibits no spectral shift; in position 3, it is receding, and the shift will be to the red. Therefore, the combined spectrum of the two stars will show variations, and the lines will be periodically doubled. Even if the spectrum of one star is too faint to be seen at all, the wobbling of the lines due to the other star will be just as tell-tale. Pairs of this kind are

Fig. 14.5. Epsilon Lyræ, the double–double star, drawn by Paul Doherty. All four main components can be seen with a 3-inch refractor.

known as spectroscopic binaries. Examples are Capella, Spica in Virgo, and both the bright components of Mizar.

We also meet with multiple stars, containing several components. One of the best-known is Epsilon Lyræ, near Vega; it is almost overhead as seen from Britain at times during summer. Keen eyes will show that Epsilon is made up of two components, and in binoculars the two can be well seen, since the separation is over 200 seconds of arc. A 3-inch telescope reveals that each component is again double, so that there are four visible stars in the field (Fig. 14.5), and to make things even more complex one of the four is itself a spectroscopic binary. The two main pairs are so far apart that they must take well over a million years to complete one revolution round their centre of gravity.

Then there is Castor, in the Twins. We have noted that it is an easy telescopic double. Each member is a spectroscopic binary, and there is a 9th-magnitude spectroscopic companion 73 seconds of arc away, so that the Castor system is made up of six separate suns – four bright, two dim. (The faint pair is actually an eclipsing binary, about which I will have more to say later. It is known either as Castor C or as YY Geminorum.)

The magnification to be used for looking at any particular double star must depend upon the individual double itself. If you want to obtain an overall view of Mizar and its companions, a low power is needed, since if you increase the magnification you will find that Alcor is out of the field. Closer pairs naturally need higher powers, and for measuring work considerable magnification must be used.

Useful research can be carried out by the amateur double-star enthusiast even if he does not possess specialized equipment, and it is true that some of the brighter pairs have been shamefully neglected. In any case, there is a great deal of enjoyment to be drawn from looking at the pairs and groups of suns. With their varied separations and their lovely contrasting colours, they are among the most beautiful of all the objects in the sky.

15

Variable Stars

Fortunately for us, our Sun is a steady, well-behaved star. It may have periods of activity when its surface is disturbed by many spot-groups and flares, but at least its total output of energy does not alter much over periods of hundreds of centuries. Other stars may not be so placid. There are some which vary in brightness from day to day, even from hour to hour, changing in both size and temperature, so that any orbiting planet would have to endure a most uncomfortable climate.

Variable star observation has become more and more important in amateur programmes during the past few years. The main reason is quite simple: variables are so numerous that with the best will in the world, professional astronomers cannot hope to keep careful track of them all – and one never quite knows what a variable star will do next. Professionals rely largely upon amateur watchers, who play their part nobly.

For really accurate measurements, some sort of electronic equipment is needed; using electronic aids together with a modest telescope, the observer can give brightness estimates down to a hundredth of a magnitude. I have never attempted this myself, and so I cannot act as a guide to others; but using eye-estimates against suitable comparison stars, I will usually back myself to make an estimate which is correct to a tenth of a magnitude, and this is good enough for most branches of research. Obviously, the larger the telescope used, the greater the number of variables which will come within range, but there are some which can be studied with the naked eye, and dozens which are easy to follow with binoculars. I have an observing list of about 60 stars, and these keep me busy on a clear night – bearing in mind that my main interests have always been within the Solar System, so that when the Moon or a planet is on view I have to let my variable stars take second place.

Variable stars are of many kinds. All I can do here is give a rough classification; I have elaborated it slightly in Appendix 25, and we must, I feel, give the basic facts before going on to say anything about methods of observation. The main types are: (1) eclipsing binaries; (2) short-period variables such as Cepheids; (3) long-period or Mira variables; (4) semi-regular variables; (5) irregular variables of various kinds; and (6) exploding stars – novæ and supernovæ.

First, then, let me consider the 'fake variables', or eclipsing binaries. They might

157

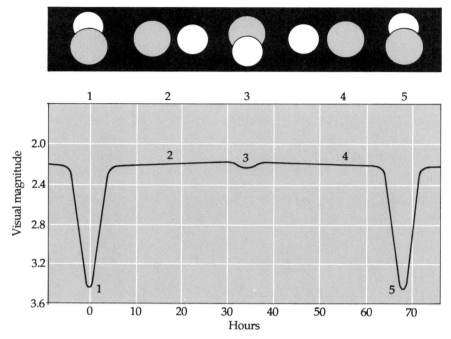

Fig. 15.1. Light-curve of Algol. At primary minimum, the smaller but brighter component is partially hidden by the fainter. The secondary minimum (position 3) is very slight.

equally well have been dealt with under the heading of double stars, but since they do seem to change in brightness they come within the scope of the variable star enthusiast.

The best-known of these 'fakes' is Algol or Beta Persei, in the constellation of Perseus; it lies in the northern hemisphere of the sky, but is visible from most inhabited countries. In mythology, Perseus was the gallant hero who slew the fearful Gorgon, Medusa, whose glance turned the hardiest onlooker to stone, and it is fitting that Algol should mark the Gorgon's severed head.

Usually Algol shines as a star of magnitude 2.1, only a little fainter than Polaris. It remains constant (or virtually so) for a period of $2\frac{1}{2}$ days, but then it starts to fade, until after about five hours it has dropped to magnitude 3.3. After staying at minimum for a mere 20 minutes, it starts to brighten once more, taking a further five hours to regain its lost lustre (Fig. 15.1). Apparently this curious behaviour was first reported by the Italian astronomer Montanari in 1667. The old Arab astronomers called Algol 'the Demon Star' and regarded it as being of evil influence – which is interesting if they were unaware of its variations.

Actually, Algol is not variable at all. It is a binary, and when the brighter star is eclipsed by the fainter the total brightness naturally drops. When the fainter star is eclipsed by the brighter, there is a secondary minimum, but with Algol the fading is so

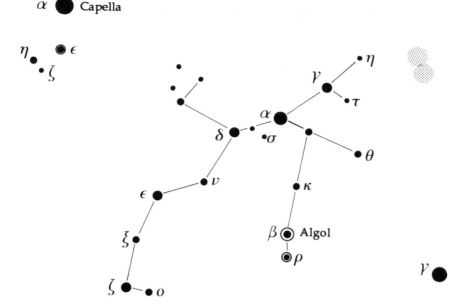

Fig. 15.2. Position of Algol, with some comparison stars; Zeta (magnitude 2.8), Epsilon (2.9), Nu (3.8) and Gamma Andromedæ (2.1). Avoid using Rho, which is itself variable.

slight that you need electronic equipment to measure it.

The brighter component is of type B8, with a diameter of about 3 times that of the Sun and a luminosity a hundred times as great; the secondary is rather larger, but cooler and less powerful. The distance between the two, centre to centre, is no more than about 7 000 000 miles, so that no telescope could show them separately even if they were equally bright. At main eclipse, about 79 per cent of the primary is covered by the secondary. The Algol system also includes a third star which does not take part in the eclipses, and there may well be a fourth component also.

Purists will note that there is an error of terminology here. Since Algol B covers Algol A, and vice versa, what we are really seeing is not an eclipse, but an occultation – and eclipsing binaries should more properly be called occulting binaries. Not that it really matters; after all, as we have seen, an eclipse of the Sun is really a lunar occultation.

With the naked eye it is easy enough to plot Algol's light-curve, by giving the magnitude against the time-lapse. The times of primary minima are given in yearly handbooks, so that you will know when to expect them; observe every few minutes, and estimate Algol against suitable non-variable comparison stars, such as Zeta (magnitude 2.8), Epsilon (2.9), Nu (3.8) and Gamma Andromedæ (2.1). Avoid using Rho Persei, which is itself variable – the sort of trap about which I will have more to say later. The Algol region is shown in Fig. 15.2. Other Algol-type eclipsing binaries are Lambda Tauri, in the Bull; Delta Libræ, in the Balance; and Zeta Phœnicis, in the far-

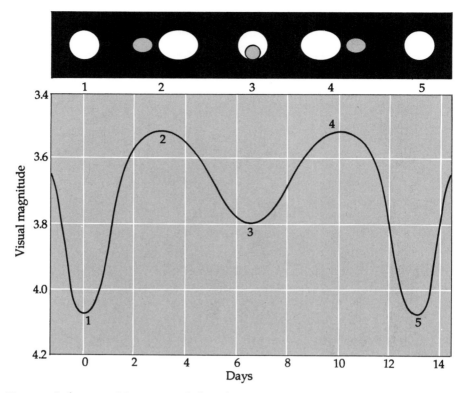

Fig. 15.3. Light-curve of Beta Lyræ, which is always in variation. There are alternate deep and shallow minima.

southern constellation of the Peacock. All these are within naked-eye range, though Delta Libræ sinks to the sixth magnitude at minimum and will then need to be observed with binoculars or a wide-field telescope.

Another eclipsing binary, but of different type, is Beta Lyræ, near Vega — sometimes still known by its old proper name of Sheliak (Fig. 15.3). Here there are two bright, somewhat unequal components, so close together that they must be almost touching each other. At maximum, when both stars are shining together, Beta Lyræ appears of magnitude 3.4. It then fades to 3.8, rises again to 3.4, and at the next minimum drops to 4.4, so that deep and shallow minima take place alternately. The brightness is always changing, so that there is no long, almost flat maximum, as with Algol. Apparently the components of Beta Lyræ are egg-shaped, because they are distorting each other gravitationally, and there seem to be streamers and clouds of gas surrounding the whole system. Seen from close range, it would be a remarkable sight. Many other Beta Lyræ stars are known, and at least two (u Herculis and the southern V Puppis) are within naked-eye range.

Two of the most remarkable of all eclipsing binaries lie close together in the sky, close to Capella in Auriga (the Charioteer); they are two members of the little triangle of stars which are known collectively as the Hædi, or Kids. Their apparent closeness is not significant. The star at the apex of the triangle, Epsilon Aurigæ, is much more remote than the other eclipsing star, Zeta Aurigæ, or the non-variable member of the trio, Eta Aurigæ.

Epsilon Aurigæ is a real puzzle. Normally it is of magnitude 3.0, and the primary — the only visible component — is an exceptionally luminous F-type supergiant, over 4500 light-years away. Every 27 years it is eclipsed by an invisible component which has never been seen at all. The eclipse is a leisurely affair; the last one began on 22 July 1982, was total between 11 January 1983 and 16 January 1984, and did not end finally until 25 June 1984, so that it extended over almost two years. Near mid-eclipse I estimated the magnitude as only 3.9, so that Epsilon was then much fainter than Eta, which is steady at magnitude 3.2.

To produce an eclipse of this duration, the companion must be very large indeed. It was once thought to be the largest star known, with a diameter great enough to swallow up the orbits of all the Sun's planets out almost as far as Saturn; then it was assumed to be a black hole; now it is more generally regarded as a small, hot star surrounded by a gaseous opaque cloud — but we do not really know, and Epsilon Aurigæ remains in a class of its own. The next eclipse is not due until the year 2009, but Epsilon is always worth watching, because it may not be quite constant even it is not officially performing.

Zeta Aurigæ, or Sadatoni, has a much shorter period (972 days) and a much smaller range (magnitude 4.9 to 5.5). The primary is a K-type supergiant, while the secondary is a hot white star of class B. Zeta is particularly interesting to spectroscopists because the smaller star can be traced for some time though the diffuse outer layers of the giant before being totally eclipsed. When an eclipse occurs, it extends over a period of about five weeks.

There are many faint eclipsing binaries of various types — YY Geminorum, the faintest member of the Castor system, is one — and they tend to be rather neglected, so that amateurs are starting to show renewed interest in them. Often the periods are short, so that observations have to be made every hour or so instead of just nightly.

Turning now to genuine variables, we must begin with the Cepheids, which are of unique importance because they are obliging enough to act as 'standard candles' (Fig. 15.4). Several are within naked-eye range, the best known being Delta Cephei itself, which lies fairly close to the north celestial pole (Appendix 28: Map VII). The period is 5.4 days, with a range of from 3.5 to 4.4, and the light-curve is not symmetrical. The rise from minimum to maximum is quicker than the subsequent fall, and this is always the case, because Delta Cephei's variations are absolutely regular. We always know how bright it will be at any particular moment, though this is not true for all Cepheids of somewhat different type.

A Cepheid is a pulsating star, expanding and contracting. This can be proved by spectroscopic observations. When a Cepheid is expanding, its bright surface is moving toward us, and the spectral lines are Doppler-shifted toward the blue; when the star is contracting, the surface is receding, and the shift is toward the red, though allowance must naturally be made for the overall radial velocity of the star (Delta

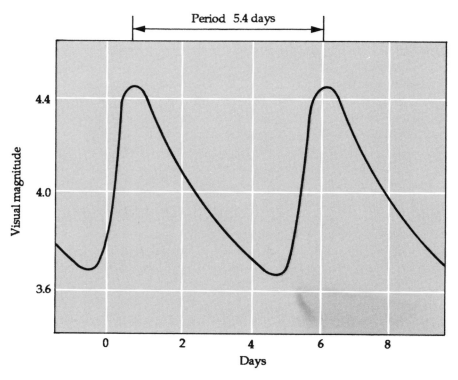

Fig. 15.4. Light-curve of Delta Cephei, which is perfectly regular.

Cephei itself is approaching us at 13 miles per second). Most classical Cepheids have ranges of no more than a couple of magnitudes. They are F-, G- or K-type supergiants, well advanced in their evolutionary cycles. In some cases the range is much less; Polaris is actually a Cepheid, but does not vary by more than a tenth of a magnitude, so that with the naked eye it appears constant.

The importance of Cepheids lies in what is called the Period–Luminosity Law. Reduced to its simplest terms, this Law links the variation period of a Cepheid with the star's real luminosity, so that variables of equal period have the same candle-power. Delta Cephei, period 5.4 days, is around 6000 times as luminous as the Sun; therefore, every other classical Cepheid with a period of 5.4 days must also be 6000 Sun-power. We can then work out the distance, which for Delta Cephei proves to be of the order of 1400 light-years. The longer the period, the more powerful the star; thus Eta Aquilæ, with a period of 7.2 days, is more luminous than Delta Cephei.

All this means that we are provided with an excellent way of measuring the distance of a remote star-cluster or galaxy. If we can detect a Cepheid, we can find its distance, and thus the distance of the cluster in which it lies. Nature can be awkward at times, as we know to our cost, but in this case she has given us an unexpectedly accurate measuring-rod. Inevitably there are complications, because we now know that there are sub-classes of short-period variables which follow different rules; there are, for instance, the Type II Cepheids or W Virginis stars, of which the brightest example is the far-south Kappa Pavonis, in the Peacock (Appendix 28: Map XVI). Kappa Pavonis has a range of from magnitude 3.9 to 5.5, and a period of just over 9 days, but it is much less luminous than a classical Cepheid of the same period would be.

There are, too, the RR Lyræ stars, with small amplitudes and periods which are usually less than a day. Many, though not all, lie in star-clusters, and they were once often called cluster-Cepheids. All of them seem to have the same luminosity, around 90 times that of the Sun, so that they also can be used as 'standard candles', but most of them are too faint to be of interest to the average amateur observer who is not equipped with electronic aids.

Variable stars of long period are quite unlike the Cepheids, and there are many within the range of small telescopes, though in some cases they become inconveniently faint at minimum.

In August 1596 the Dutch astronomer David Fabricius recorded a third-magnitude star in Cetus, the Whale (Appendix 28: Map X). By October it has disappeared. Johann Bayer saw it again in 1603, when he was compiling his star-catalogue, and gave it the Greek letter Omicron, but shortly afterwards it vanished once more. Not until 1638 was if found that it appears with fair regularity; it takes approximately 332 days to pass from one maximum to the next, and it is visible with the naked eye for only a few weeks in every year. It was named Mira, or 'the Wonderful', and long-period variables are now officially known as Mira stars (Fig. 15.5).

The period of naked-eye visibility is not always the same, and neither is the magnitude at maximum. In 1969, and again in 1987, it reached almost the second

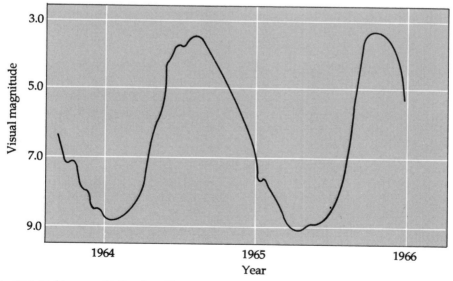

Fig. 15.5. Light-curve of Mira Ceti. The mean period is 332 days, but both the period and the amplitude are themselves subject to variations. This shows the curve from 1964 to 1966; in 1987 Mira became as bright as magnitude 2.3.

magnitude; on 18 February 1987 I made it 2.3, almost the equal of Polaris, and records dating back to 1779 indicate that in that year it became even brighter, rising to the first magnitude. At other maxima it barely exceeds magnitude 4. The period also is inconstant; it can range between 304 and 355 days, so that 332 days is only an average. There is nothing neat or precise about Mira, and it is always worth keeping under watch. At minimum the magnitude falls to 10, so that Mira cannot then be seen even with binoculars, though a 3-inch telescope will always be adequate to follow it. There is a faint binary companion which seems to be a hot sub-dwarf; Mira itself, like all long-period variables of its type, is a huge, diffuse red giant. The spectral type is M, and the distance seems to be about 95 light-years.

Mira stars are plentiful enough; other naked-eye examples at maximum are Chi Cygni in the Swan and R Hydræ in the Watersnake, but generally speaking telescopes are needed to study them. Chi Cygni is of special interest. It has a period of 407 days, and although it is quite bright at its best – in September 1987 it rose to magnitude 4.5 according to my observations, and it has been known to become even more conspicuous – it sinks to below 14 at minimum, and it is then decidedly tricky to identify, particularly since it lies in a rich field. It has an S-type spectrum, and is intensely red. It was identified as being variable in 1686, and was actually the third variable star to be discovered, following Mira (1638) and Algol (1669).

The long-period stars show no period-luminosity law, and so cannot be used for distance-measuring. Like the Cepheids, they are pulsating, but in general they are much less powerful. R Hydræ, with a range of magnitude from 4.0 to 10.0, is remarkable inasmuch as its period has changed markedly during the past couple of hundred years; apparently it used to be almost 500 days on average, but is now no more than 386, though again it is not constant in its behaviour.

Loosely associated with the long-period variables are the semi-regular stars, which are in general red giants or supergiants; they have smaller amplitudes, and their fluctuations are erratic, though there is enough periodicity to be recognizable – at least in most cases. Betelgeuse, in Orion, is the most famous example. The extreme range seems to be from about magnitude 0.1 to 0.8, so that at maximum there may be almost no difference between Betelgeuse and Rigel, while at minimum Betelgeuse is little brighter than Aldebaran in the Bull. It is said that the period is 2070 days, just over $5\frac{1}{2}$ years, but I admit to being somewhat sceptical. Other prominent semi-regulars are Mu Cephei (3.6–5; officially 730 days) and Rasalgethi or Alpha Herculis (3–4; 130 days). All these are of spectral type M, but Mu Cephei is particularly red, and it is easy to understand why Sir William Herschel nicknamed it 'the Garnet Star'. The colour is not very obvious with the naked eye, because the star is never bright enough, but binoculars bring it out splendidly. Actually, Mu Cephei is much more luminous than Betelgeuse, and may be the equal of at least 50 000 Suns, but it is over 1300 light-years away. If it were as close to us as (say) Sirius, it would cast shadows.

In my view, the most fascinating of all variable stars are those which are absolutely unpredictable. Gamma Cassiopeiæ, the middle star of the W of Cassiopeia, is hot, white and unstable; apparently it periodically throws off shells of material. Usually it is of just below the second magnitude, but in 1936 it brightened up to 1.6; by 1940 it had

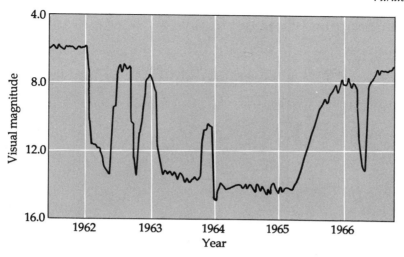

Fig. 15.6. Light-curve of R Coronæ Borealis. This is for the period 1962–66, when there were several minima. The behaviour of R Coronæ stars is always unpredictable.

dropped to below 3, and then slowly recovered, though it is never quite steady, and may surprise us again at any time. It is a very suitable star for naked-eye observers, since it lies close to the constant Beta (2.3). It also lies to the orange, K-type Shedir, or Alpha Cassiopeiæ, which is itself suspected of being variable over a small range – perhaps 2.1–2.4, according to my estimates. Close to Beta, and in the same binocular field, is the interesting Rho Cassiopeiæ, which is usually of around the fifth magnitude, but has been known to experience occasional falls to below 6. This has not happened now for almost 40 years – but – who knows? As yet we are not sure whether or not it can be fitted into any definite class. I have been waiting for it to 'perform' ever since I began observing it in 1965, but so far it has not obliged.

Telescopic irregular variables are of many kinds. R Coronæ Borealis, in the Northern Crown, is generally of about the 6th magnitude, on the fringe of naked-eye visibility, but at irregular intervals it fades sharply, dropping to below magnitude 14 and taking weeks or months to recover (Fig. 15.6). It is a luminous G-type supergiant, and apparently fades when clouds of nothing more nor less than soot accumulate in its atmosphere, blocking out part of the light coming from below. R Coronæ stars are rare, but there are a few within reasonable range – for example SU Tauri, in the Bull, and RY Sagittarii, in the Archer – and they should be carefully watched, because the onset of their minima cannot be predicted. On the contrary, stars of the SS Cygni or U Geminorum type remain at minimum for most of the time, but periodically flare up by several magnitudes; with SS Cygni itself, the brightest member of the class, outbursts occur around every 50 days, when the magnitude increases quite quickly from below 12 to as high as 8.1 (Fig. 15.7). All SS Cygni stars are binaries, made up of a K or M dwarf together with a white dwarf; the outbursts occur when the white dwarf has pulled enough material away from its larger, less dense companion to create

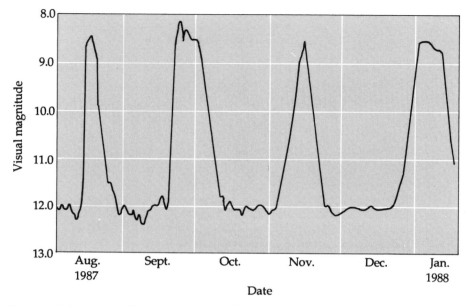

Fig. 15.7. Light-curve of SS Cygni, 1987–88. This curve was drawn from my observations with 15-inch and 12½-inch reflectors.

conditions of instability.* RV Tauri variables are pulsating F to K-type supergiants with roughly alternate deep and shallow minima; one of them – R Scuti, in the Shield – can be visible with the naked eye at maximum, and is always within the range of good binoculars. The most erratic of all variable stars must surely be Eta Carinæ, in the Keel, which is unfortunately too far south to be visible from Europe or most of the United States. For some years, between the mid-1830s and the mid-1840s, it was the brightest star in the entire sky apart from Sirius, but for a century now it has been invisible with the naked eye, though it may increase again at any moment. Eta Carinæ is unique; it is associated with nebulosity, and through a telescope it looks to me more like an 'orange blob' than an ordinary star. At its peak it may have been at least 6 000 000 times more powerful than the Sun, making it the most luminous star known. Even today it is almost as energetic, though at the moment much of its radiation lies in the infra-red part of the spectrum. It is exceptionally massive, and almost certainly unstable.

Variable star estimates are made visually by comparing the variable with suitable comparison stars which do not alter. One difficulty of observing naked-eye variables such as Betelgeuse is that a star is bound to be reduced in brightness when near the horizon, since it will be shining through a thicker layer of the Earth's atmosphere. This 'extinction' effect can upset an observation completely if it is not allowed for, but the table given in Appendix 25 should help. With telescopic observations, extinction can

---

* Stars of this type are generally known as SS Cygni variables in America, U Geminorum variables in Europe. U Geminorum has a range of from magnitude 8.8 to 14.4; the mean interval between outbursts is 103 days.

be neglected, because all the stars in the field will be at approximately the same altitude above the horizon.

The first thing to do is to identify the variable. A star atlas is essential, together with a chart of the type similar to those in Appendix 25, and the position of the variable can be found – but take care! It is often a mistake to look directly for the variable itself. The best method is to check the stars which lie in the same low-power field, and build up an overall picture. Most Mira stars are distinguishable because of their colour, but this is never a safe guide.

It may sound difficult to identify any particular star-field, but no two fields are exactly alike, and a little practice will work wonders. Aimless 'sweeping around' is emphatically not to be recommended. First identify the area by the naked-eye stars which can be recognized without possibility of error, and then proceed by means of star alignments and patterns, swinging the telescope north or south and in right ascension in terms of a known angular field. In troublesome cases, an easily recognizable star can be selected which has the same declination as the variable, and the telescope left stationary until the variable comes into view – though admittedly this tends to waste time, and is not practicable unless you have a telescope on an equatorial mount. Never leave any 'safe anchorage' for the next until you are quite certain that you have made no mistake. Once you have found a field, you will generally be able to recognize it again without much trouble; even so, the approach should always be systematic. A moment's carelessness can lead to some very peculiar results.

Obviously some fields are more obliging than others. Consider, for example, the long-period Mira star U Orionis and the much fainter R Coronæ type star SU Tauri. Here, the trick is to begin with the naked-eye Chi¹ and Chi² Orionis, which you can locate without any trouble at all. Put them in the field of the finder, and you will have U Orionis in view; it has a range of from magnitude 5.3 to 12.6, and a period of 372 days. Now swing across to a curved arrangement of stars; to me it is almost as if they were saying 'This way to SU Tauri!' and if you follow them along, you will come to the characteristic little pattern which includes the variable. Whether you will always see SU Tauri is another matter, because although it is usually between magnitudes 9 and 10 it can fall to 16 at minimum, and a large telescope is then needed to show it at all.

The magnification to be used again depends on the variable and its field. If the star is very faint, you will need a high power, but there are also some faint, irritating variables when any adequate magnification puts all the comparison stars out of the field.

There are two main methods of estimation. One of these is Pogson's Step Method, when the observer trains himself to gauge a difference of 0.1 magnitude, or one 'step'. Suppose that you are observing a variable star, and find that it is two steps fainter than comparison star A and one step brighter than comparison star B. Look up the magnitudes of A and B; if A is 8.0 and B is 8.3, the variable must be 8.2. But never trust to only one comparison star, if you can help it. Select several, some brighter than your target and some fainter. If the results are concordant, all is well; if not, there is something wrong. A discrepancy of a tenth of a magnitude can be tolerated, but not much more.

I have always used Pogson's method, but most observers seem to prefer the Fractional, which is said to be slightly more precise. Two comparison stars are used, and the brightness difference between them is divided mentally into a convenient number of parts, after which the variable is placed in its appropriate position in the scale. Suppose that you have two comparison stars, A (magnitude 7.0) and B (7.6), and that you divide the difference into fifths. If the variable is $\frac{3}{5}$ of the way from A to B, you write A 3 V 2 B, and the magnitude works out at 7.36, which you 'round down' to 7.4. Then take two more comparison stars, repeat the process, and see whether the result is the same.

Experience is all-important in variable-star work, but there are various traps which can be avoided if you are alive to them. First, go to your telescope with an open mind. If you expect a variable to be of (say) magnitude 7.5, there is a strong chance that you will in fact record it as 7.5, whether this is correct or not! Secondly, remember that when you are looking at two equal stars in the same telescopic field, the lower one will generally appear the brighter. More serious is the Purkinje effect, which becomes noticeable when you are trying to compare a red star with a white one. If the two are seen as equal, and then are brightened, the red star will apparently increase more than the white; if they are reduced by the same amount, the red star will become the fainter. There is no real answer to this, but avoid staring at the red star for too long. Some observers actually put the stars out of focus, claiming that this leads to more reliable estimates, but it is not a procedure which I would personally recommend. Finally, beware of false fields and − above all! − of variable comparison stars.

Flare stars are dwarfs which may suddenly brighten appreciably in a matter of a minute or so, presumably because of intense activity in their chromospheres. UV Ceti, in the Whale, is the classic example; its usual magnitude is not much above 13, but it has been known to rise abruptly to 5.9. Flare-star observation is tedious but rewarding; obviously it means working with a team.

The so-called 'secular variables' are stars which seem to have slowly brightened or faded over the course of centuries. For instance, Megrez, in the Plough or Dipper, was said by some old observers to be equal to the other stars of the pattern, whereas it is now a magnitude fainter; Acamar or Theta Eridani, in the River, and Denebola in Leo are other stars reported to have beome dimmer, whereas Alhena in Gemini was originally said to be of the third magnitude and is now ranked as 1.9. Frankly I am sceptical about most of these alleged changes, but one never knows, and the stars concerned are worth watching; Megrez at least has been suspected of slight variability in modern times.

Next we must turn to 'temporary stars' or novæ. Occasionally a star will blaze up without warning; it may become very bright for a few days, weeks or months before fading back to obscurity. The word 'nova' means 'new', but a nova is not really a new star at all. It originates in a binary system, where one component is a normal star and the other is a white dwarf. The white dwarf pulls material away from its companion, and this builds up until nuclear reactions are triggered off; the result is a temendous outburst which takes a long time to die down. Obviously there is some similarity with the behaviour of an SS Cygni variable, but with a nova the whole scale is immeasurably greater. Some novæ develop gaseous 'shells' which gradually expand and fade away.

Telescopic novæ are not too uncommon, and there are a few novæ which become brilliant naked-eye objects. Such were Nova Persei 1901 and Nova Aquilæ 1918, which reached magnitudes 0.0 and − 1.1, respectively − so that for a brief period Nova Aquilæ shone more brightly than any star in the sky apart from Sirius. Both are still visible, but have become very dim. DQ Herculis, the nova of 1934, was discovered on 13 December by an amateur astronomer, J.P.M. Prentice, who had been observing meteors and was taking a nocturnal stroll before retiring to bed. It rose to magnitude 1.2, and had an unusually long maximum; as it faded it developed a strong greenish hue, which was striking with the 3-inch refractor I had just acquired.

One very interesting nova, HR Delphini, was discovered in 1967 by the British amateur George Alcock, who uses special mounted binoculars. It was then of between the fourth and fifth magnitudes. I was among the first to confirm it − though I can claim absolutely no credit, because George Alcock telephoned me in the dead of night and told me exactly where it was. It brightened very slowly to above the fourth magnitude, and the subsequent fading was gradual; in early 1989 it was still above the 13th magnitude, roughly the same as it had been before its outburst. HR Delphini is the 'slowest' nova on record, and I doubt whether it will fade much further.

On the evening of 29 August 1975, I went outdoors soon after dark to make some routine observations. As soon as I looked up at the 'cross' of Cygnus, I realized that there was something wrong with it. A new star had been added, not far from Deneb; it was of magnitude 2.4, and could not possibly be overlooked, but equally certainly it had not been there on the previous night. As soon as I was satisfied that there was no mistake, and that I was not being tricked by a slowly-moving artificial satellite, I reported it. As I expected, I had been forestalled; Japanese observers had discovered it some hours earlier, before nightfall in Britain. Actually I think that I was the eight-third discoverer in order of priority!

Nova Cygni reached magnitude 1.8 on the following night, but its decline was swift. I made the magnitude 3.2 on 1 September, 4.3 on the 2nd and 5.1 on the 4th; by the 7th of the month it was down to 6.4, below naked-eye visibility, but was about the reddest star I have ever seen. The decline continued, and by February 1976 was well below 10. I have long since lost sight of it with my telescope, and so have most other observers. If HR Delphini was a particularly leisurely nova, then Nova Cygni was particularly rapid (Fig. 15.8).

Novæ generally appear near the Milky Way zone, but they are quite unpredictable, and the increase in light is usually so quick that the amateur sky-watcher has an excellent chance of making a discovery − provided that he has enough patience. Chance may play a part, as it did with Prentice's detection of DQ Herculis or, for that matter, my own independent sighting of Nova Cygni, but there are some amateurs who make systematic searches. Alcock is one. He has memorized the positions and magnitudes of at least 30 000 stars, so that as he sweeps around with his binoculars he

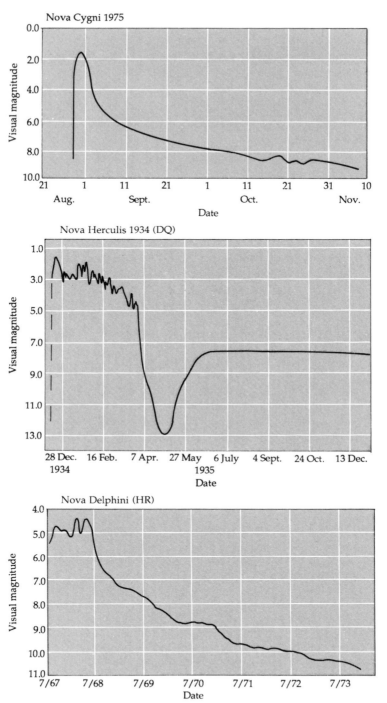

Fig. 15.8. Light-curves of three novæ: (Top) Nova (V 1500) Cygni 1975 — a fast nova, which rose abruptly to maximum and faded quickly; (centre) Nova DQ Herculis 1934; a slow nova; (bottom) HR Delphini, 1967, the slowest of all novæ; this curve has been drawn from my observations. It is unlikely that the nova will fade much further.

can detect anything unusual straight away. The Japanese are particularly good at nova-hunting, and it is not often that a newcomer can slip through their observing net. On the other hand, take care, and check on asteroids. I once firmly believed that I had discovered a telescopic nova, but I had the sense to consult my tables before reporting it. My 'nova' proved to be Asteroid No. 5, Astræa, which had been known since the year 1845.

Normal novæ may be spectacular, but they pale into insignificance compared with supernovæ, which are the most colossal explosions in all Nature. At maximum, a supernova may shine as brilliantly as all the other stars in its galaxy put together — thousands of millions of times more powerfully than the Sun. They are of two classes. A Type I supernova involves the complete destruction of the white dwarf component of a binary pair; the dwarf literally blows itself to pieces. A Type II supernova indicates the death of a more massive star which runs out of nuclear fuel, ending up as a cloud of gas in rapid expansion together with a quickly-spinning neutron star remnant which may send out pulsed radio emissions.

Four supernovæ have been seen in our Galaxy during the past thousand years, but all of these outbursts were in pre-telescopic times; the last was Kepler's Star of 1604. Of the others, the 1054 supernova, near Zeta Tauri in the Bull, has left the remnant we now call the Crab Nebula, in the centre of which is a pulsar — one of the few pulsars to have been detected optically as well as at radio wavelengths. Astronomers would dearly like to have the chance of studying a galactic supernova with modern equipment. So far they have been disappointed, but in February 1987 came the next best thing — a supernova in the Large Cloud of Magellan, a galaxy which is a mere 170 000 light-years away. The supernova — known officially as SN 1987A — took everyone by surprise (Fig. 15.9). Its declination was around 70 degrees south, so that it could not be seen from anywhere in Europe or the mainland United States (it just appeared above the horizon from the top of Mauna Kea, in Hawaii), but all southern telescopes were promptly diverted from other programmes to take advantage of it.

When first seen, the supernova was of the fourth magnitude, but instead of fading it slowly brightened up. When I observed it, on May 21 (from Johannesburg) it was of magnitude 2.3, and completely dominated the area; in colour it was orange, very like a star of type K. Subsequently it faded, but from the outset it was clear that it was a very peculiar object indeed. There is no doubt that the outburst occurred in a previously-known star which had been given the catalogue number of Sanduleak − 69°202 − but this was a blue supergiant, not a red one, so that obviously some of our theories about late stages in stellar evolution will have to be revised. Also, SN1987A was under-luminous by supernova standards, even though it was still a conspicuous naked-eye object across such a vast distance. It is the only supernova to have become visible without optical aid since Kepler's Star, though an outburst seen in 1885 in the Andromeda Galaxy, more than two million light-years away, approached the sixth magnitude.

Because supernovæ are so luminous, they can be observed in very remote galaxies, and in recent times there are a few amateurs who have made special searches for them. Pride of place must go to an Australian, the Rev Robert Evans, who uses a 12-inch

Fig. 15.9. The supernova in the Large Cloud of Magellan, photographed by Akiro Fujii from Australia on May 3 1987. The supernova was then at maximum.

reflector. He has memorized the positions and appearances of many galaxies, and during the course of a night's observing he swings from one to the other, looking carefully for anything unusual. His present tally is more than a dozen – a quite remarkable record. Others 'computerize' their telescopes, so that when the night's programme is started the telescope will function on its own, surveying the target galaxies in turn and taking photographs of them. The age when the amateur confined himself almost exclusively to the Moon and planets is well and truly over.

I came to variable star observation comparatively late in my observing career, but it is a branch of research which I strongly recommend. At the worst, you can carry out useful routine work; at best, you may emulate Robert Evans or George Alcock, and make a really dramatic discovery. I wish you luck.

16

Star-Clusters and Nebulæ

Some way from Orion, beyond the bright red star Aldebaran in Taurus, you can see what at first sight looks like a misty patch. Look more closely, and you will see that it is made up of stars. If you have average eyes, you will be able to make out seven stars under good conditions – hence the cluster is popularly called the Seven Sisters, though its official name is the Pleiades (Fig. 16.1). Keen-sighted people can do better, and the record is said to be nineteen, though even low-powered binoculars will show dozens, and the whole cluster includes hundreds of members.

The Pleiades cluster is a genuine group (it has been calculated that the odds against the chance alignment of the seven brightest stars are millions to one against), and is typical of what is known as an open or loose cluster, of which there are many in our Galaxy. Some others are visible with the naked eye – for instance the Hyades, round Aldebaran; Præsepe or the 'Beehive', in Cancer; the Sword-Handle of Perseus and, in the far south, the glorious 'Jewel Box' in the Southern Cross.

A pair of binoculars will show the Pleiades very well, and in fact the view is better than in a telescope, because unless the telescopic field is very wide only part of the cluster will be seen at any one time. The Pleiad stars look close together, but there is no fear of collision. Neither must we be deceived by the fact that the whole cluster looks so small; its real diameter is over 50 light-years. Its distance from the Solar System is 410 light-years.

The Hyades, round Aldebaran, are not so spectacular; the stars are wider apart, making up a sort of V-formation which is easy to see with the naked eye. Aldebaran itself is not a true member of the cluster, and merely happens to lie about midway between the Hyades and ourselves. On the whole this is rather a pity, since the fainter stars are over-powered by the bright orange–red light of Aldebaran.

There is another important difference between the two clusters. In the Pleiades, the brightest stars are blue, highly luminous and relatively young; there is also scattered nebulosity, not easy to see with a telescope but by no means hard to photograph. In the Hyades, the leading members are orange, and of type K. Bluish and white stars do occur, but they are much less in evidence, so that the Hyades are obviously more advanced in their life-stories.

Præsepe, in Cancer (the Crab) is nicknamed the 'Beehive'; it is easy to find, about

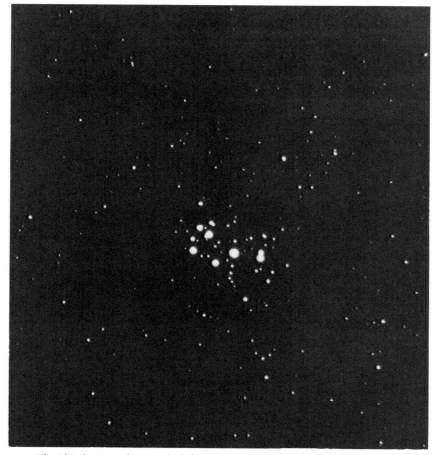

Fig. 16.1. The Pleiades, as I photographed them with an ordinary camera.

half-way between Regulus and Pollux, and flanked to either side by fainter stars known as the Aselli or 'Asses' (Fig. 16.2). Here again a low power is recommended, and the best view is probably obtained with binoculars. In Perseus we find two clusters in the same telescopic field, provided that the magnification is low; with the naked eye they can be made out as a very dim, misty patch. And in Crux, the Southern Cross, we have the cluster round Kappa Crucis, nicknamed the 'Jewel Box'; most of its stars are bluish-white, but one is red, and stands out magnificently.

These are only a few examples of open clusters; if you have binoculars, or almost any telescope, you will be able to find dozens, each of which has its own characteristics. Most of the brightest are contained in the 1781 catalogue drawn up by Charles Messier during his comet searches; thus the Pleiades cluster is M.45; the open cluster on Canis Major is M.41 (Fig. 16.3), and so on. We also have N.G.C. numbers, allotted by John Dreyer in his New General Catalogue of Clusters and Nebulæ — though by now it is hardly new; it was compiled a century ago.

Messier's list also includes some globular clusters, which are symmetrical systems containing up to a million stars. Near their centres, the stars are so close together in the

Fig. 16.2. M.44 (Præsepe).

sky that they merge, though once again appearances are deceptive; the stars even in the middle of a globular are many millions of miles apart. All the globulars are very remote. They form a sort of 'outer surround' to the main Galaxy; and since we are not in the middle of the Galaxy, we have a somewhat lop-sided view. Most of the globular clusters are in the southern hemisphere, particularly in the area of Scorpius and Sagittarius. We know their distances accurately, because they contain short-period variables which serve as 'standard candles'.

Because they are so far away, most of the globular clusters look faint, and only three are clearly visible with the naked eye. Incomparably the finest are Omega Centauri and 47 Tucanæ, both of which are too far south to be seen from Britain, though Omega Centauri does rise from parts of the United States (Fig. 16.4).

Omega, particularly, is superb. Use binoculars, and you will see stars in its outer regions; with a telescope, it seems that you can resolve the cluster almost to its centre. 47 Tucanæ is similar, and in a way it is even more striking, because it is smaller and can therefore be fitted into a smaller field with higher magnification. In the northern sky we have to make do with Messier 13, in Hercules, which you can find between the stars Zeta and Eta Herculis, rather closer to Eta. Without claiming for a moment that it can rival the southern pair, it is still a wonderful sight in a telescope of larger than 3-inches aperture. Fainter globulars naturally require more powerful instruments, but they are well worth seeking out.

Messier catalogued only a little over 100 objects of various kinds. Herschel added many more, and listed them; he found that while some were obviously 'starry', others were not, and looked like shining gas. These were the nebulæ, from the Latin for 'clouds'.

Fig. 16.3. M.41, a prominent open cluster in Canis Major not far from Sirius. Photographed with a telephoto lens on a camera mounted on my 15-inch reflector.

Were the nebulæ really nothing more than star-clusters, so far away that they could not be resolved? There were also the inappropriately-named planetary nebulæ, which showed pale disks. The problem was solved in 1864 by Sir William Huggins, one of the great pioneers of astronomical spectroscopy, when he turned his attention to a planetary nebula in Draco, the Dragon. He half expected to see a somewhat confused effect due to the combined spectra of thousands of stars; instead, he saw nothing except a single green line. The light of the nebula was made up of one colour only, emitted by luminous gas. We now know that a planetary nebula consists of a faint, hot star associated with an immense gaseous 'shell' which is expanding. The central star is all that remains of a sun which has left the Main Sequence and has puffed off its outer 'atmosphere'.

Fig. 16.4. The globular cluster Omega Centauri — much the brightest in the sky, but in the far south. (Photograph by A. Page.)

The best-known planetary nebula is M.57, the Ring Nebula in Lyra, close to Vega; it is easy to find, because it lies between two fairly bright stars, Beta Lyræ (the celebrated eclipsing binary) and its third-magnitude neighbour, Gamma Lyræ. A 3-inch telescope will show it as a very faint, luminous ring, but a much larger aperture is needed to bring out the central star; I can see it with my 15-inch reflector, and no doubt smaller telescopes will show it to keen-eyed observers. Other planetaries, such as the Dumbbell in Vulpecula (the Fox) are less symmetrical (Fig. 16.5). In general they are somewhat elusive, because of their low surface brightness, but look carefully for M.97, the Owl Nebula in Ursa Major; I have found it with a 6-inch reflector, but at least a 12-inch is needed for a really good view.

A planetary nebula is a dying star; a diffuse nebula is a stellar nursery, in which new stars are being formed from dust and gas. The best-known example is M.42, the Sword of Orion, which is easily visible with the naked eye close to the three stars of the Hunter's Belt (below the Belt as seen from Britain, above the Belt from Australia or South Africa) (Fig. 16.6). It is one of the show-pieces of the sky, particularly as it contains the multiple star Theta Orionis, whose four main components make up the Trapezium. M.42 is around 1500 light-years away, and makes up part of a vast 'molecular cloud' which covers most of Orion.

It is the Trapezium which makes the nebula shine — partly by reflection, partly because the stars are so hot that they excite the gas to a certain amount of self-luminosity. Not that the gas is dense; it is millions of times more rarefied than the air that you and I are breathing.

Fig. 16.5. The Dumbbell Nebula, M.27; (Photograph by Ron Arbour, 40 cm reflector, Tri-X, exposure 8 minutes.

Fig. 16.6. The Great Nebula in Orion, M.42. (Photograph by Ron Arbour, 12 January 1988, 40 cm reflector, exposure 20 minutes.)

Fig. 16.7. The Horse's Head Nebula in Orion. (Photograph by R. W. Arbour, 12 January, 1989.)
The bright over-exposed star in Alnitak.

There are other diffuse nebulæ prominent enough to be made out with small or moderate telescopes — for example M.20 (the Trifid Nebula) and M.8 (the Lagoon Nebula), both in Sagittarius; but it must be said that not much detail can be seen visually, though photographs of them can be very rewarding.

If a nebula includes no suitable stars to illuminate it, it will remain dark, and will be detectable only because it will blot out the light of stars beyond. William Herschel was inclined to believe that the occasional well-defined, virtually starless patches were true 'holes in the heavens', but we now know that there is no basic difference between a nebula which shines and one which does not. Of the dark nebulæ, the most obvious is the Coal Sack in the Southern Cross; there are others elsewhere, notably in Cygnus, not far from Deneb. In Orion we find the Horse's Head, which really does give the impression of a knight's head in chess (Fig. 16.7). It is not hard to photograph, with a reasonable aperture and a good clock-drive, but it is remarkably difficult to see visually.

It is perhaps fitting that the first entry in Messier's catalogue should be the Crab Nebula, the remnant of the supernova of 1054 (Fig. 16.8). Good binoculars will show it; with a telescope it appears as a faint blur; photographs taken with powerful instruments reveal its incredibly complex form. It sends out radio waves, X-rays, gamma-rays and almost everything else; it contains a pulsar, and it is one of the most interesting objects in the sky, dim though it may seem. Look for it by all means, but do not expect anything spectacular.

Fig. 16.8. The Crab Nebula. (Photography by R. W. Arbour.)

In general, a low power is to be recommended for observing nebulæ, apart from the planetaries; a magnification of ×30 or ×40 will do quite well. It is also worth remembering that more than five thousand million years ago, our own Sun was born inside a diffuse nebula.

17

Galaxies — and the Universe

One of the glories of the night sky is the luminous band which we call the Milky Way. It stretches from one horizon to the other, from Cassiopeia in the far north to the Southern Cross in the far south, and on a clear, moonless night it is truly magnificent.

It was Galileo, in 1610, who first realized that the Milky Way is made up of stars. He believed that there must be 'an infinite number' of them; he was — understandably — wrong, but our star-system or Galaxy contains about a hundred thousand million members, together with a vast amount of thinly-spread intersteller material.

Sweeping the Milky Way with binoculars or a wide-field telescope will show so many stars that any attempt to count them would be doomed to failure (Fig. 17.1). The band is fairly well defined, though in some places it is much richer than in others. You may well imagine that the stars in it are so crowded together that they are in serious danger of collision — but as so often happens — appearances are deceptive. The Milky Way is not a crowded place, and the luminous band is nothing more than a line of sight effect, due to the shape of the Galaxy itself. The Galaxy is flattened, with a central bulge; the position of the Sun is not far from the main plane, but well away from the galactic nucleus or centre of the system (Fig. 17.2). The overall diameter is around 100 000 light-years, and the greatest breadth is about 20 000 light-years. The distance between the Sun and the galactic nucleus is around 30 000 light-years; opinions vary, but this figure is certainly not very far wrong. As you look through the plane of our Galaxy you will see many stars in almost the same direction, producing the Milky Way band.

Actually, we cannot see all the way from the Sun to the centre, because there is too much interstellar 'dust' in the way, and starlight cannot penetrate it, any more than a car's headlights can pierce a thick fog. True, the interstellar dust is very thinly distributed, but there is an immense volume of it. The galactic centre lies in the direction of the glorious star-clouds in Sagittarius (the Archer), and is a somewhat mysterious place; we can pick up infra-red radiations from it, but it is forever hidden from our view. There are even speculations that it may contain a massive black hole.

Round the main part of the Galaxy we have the so-called galactic halo, which is more or less spherical; it contains isolated stars as well as the globular clusters. And if we could see the Galaxy in 'plan view' it would show up as being of spiral form, not

Fig. 17.1. The Milky Way. This composite picture was taken at the Lund Observatory, Sweden. The two blobs below the main Milky Way system are the Magellanic Clouds.

unlike a Catherine-wheel. It is rotating round its centre; the Sun takes 225 000 000 years to complete one orbit, a period which is sometimes nicknamed the cosmic year. One cosmic year ago the most advanced life-forms on Earth were amphibians, so that even the dinosaurs lay in the future. What will happen by the end of the next cosmic year is anybody's guess!

There are two main types of 'population' in the Galaxy. Population I stars, such as the Sun, are to be found in or near the spiral arms; the most luminous of them are blue supergiants, while the leading members of Population II are old red giants well advanced in their evolution. Population II makes up the galactic halo, and also, presumably, the region near the centre. Since Population II stars are revolving more slowly round the nucleus, and in more elliptical orbits, they seem to have high velocity relative to the Sun. Note, too, that in Population I areas star-formation is still going on, and there are many clouds and diffuse nebulæ, while Population II areas have been more or less 'swept clean', with all the star-forming material used up.

There is no point in saying much about methods of observing the Milky Way, except to repeat that a low magnification is suitable, and that the very best views will be obtained with binoculars or a very wide-field eyepiece. There are endless star-fields, together with clusters, groups and dark rifts.

I have already said something about the Clouds of Magellan in the southern sky (Fig. 17.3). Superficially they look rather like detached portions of the Milky Way; actually they are separate systems, and are often regarded as satellites of our Galaxy. The Small Cloud lies almost behind the famous globular cluster 47 Tucanæ, and it is tempting to think that the two are related, though in fact they are not. When you look at them through binoculars or a telescope, it is very noticeable that the surface brightness of the cluster is much higher than that of the Cloud.

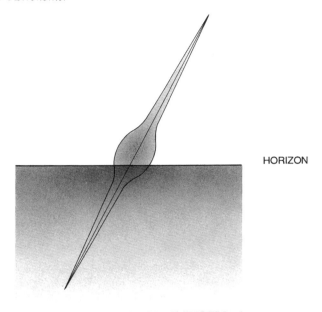

HORIZON

Fig. 17.2. Cross-section of our Galaxy, showing the position of the Sun.

Fig. 17.3. The Magellanic Clouds.

Fig. 17.4. The Andromeda Spiral, M.31 (Mount Wilson). This shows the spiral structure.

Both the Clouds are irregular in form, though it has been claimed that the Large Cloud shows signs of spiral structure while the Small Cloud may be made up of two distinct parts. The distances as given by astronomers at the Royal Greenwich Observatory are 170 000 light-years for the Large Cloud and 190 000 light-years for the Small Cloud.

The only other galaxy clearly visible with the naked eye is the northern Messier 31 (M.31), the Great Galaxy in Andromeda, which you can just make out as a dim smudge under good conditions, and was recorded by the Arab star-gazers of a thousand years ago (Fig. 17.4). Binoculars show it clearly, but large-scale photographs are needed to bring out its true form, though a smaller telescope will show it together with its two smaller companion galaxies, M.32 and NGC 205. Like our Galaxy, M.31 is spiral, but unfortunately it lies at a narrow angle to us, so that its full beauty is lost. Were it face-on — like, for instance, the Whirlpool Galaxy M.51, in the Hunting Dogs — it would be superb (Fig. 17.5).

Messier catalogued many of the so-called starry nebulæ, and Herschel and others added many more. It was only in 1845 that the Earl of Rosse, using his extraordinary 72-inch reflector at Birr Castle, found that some of them, including M.51, were spiral. Yet even then it was still thought that the resolvable nebulæ — spiral and otherwise — were members of our Galaxy, and fairly junior members at that.

Eventually the problem was resolved by the use of those convenient 'standard candles', the Cepheid variables. In 1923 Edwin Hubble, using what was then the most powerful telescope in the world, the 100-inch Mount Wilson reflector (which was temporarily closed because of light pollution from Los Angeles), was able to identify

Fig. 17.5. The Whirlpool Galaxy, M.51 in Canes Venatici. This photograph brings out the spiral structure and also the satellite galaxy. (Photograph by Ron Arbour, 40 cm reflector.)

Cepheids in the Andromeda Spiral. As soon as he measured their distances, he realized that the Spiral was much too remote to belong to our System. The modern value for its distance is 2 200 000 light-years; even so, it is one of the very nearest of the external galaxies.

Today there must be at least a thousand million galaxies within range of our telescopes. Small apertures will show dozens, and you will be able to see that they are by no means all alike; there are elliptical systems, galaxies which look rather like globular clusters, irregulars, and even the remarkable barred spirals, in which the arms seem to issue from a well-marked bar running along the main plane and through the nucleus.

Searching for galaxies is fascinating, and I know many amateurs who have worked right through the objects in Messier's list as well as many of the NGC objects. I used to think that there was not much scientific value in it, apart from sheer enjoyment, but I have changed my mind now, because of the amateur successes in supernova-hunting. Who knows? – a tremendous outburst may take place at any time in one of the outer systems, and an amateur may well be the first to report it. In 1885 we did have a supernova in the Andromeda Spiral, discovered at about the same time by several independent observers – including a Hungarian baroness who was having a house-party and had set up a portable telescope on one of the lawns in front of her castle! (It is a pity that this star, remembered today as S Andromedæ, appeared too early for us to take advantage of it. In 1885 its nature was not appreciated, and we did not even know that the Spiral itself was a separate galaxy. Its remnant was identified as a tiny patch as recently as 1989.)

Therefore, it is well worth making a list of suitable galaxies, memorizing their positions, making sure that you know just what they look like, and surveying them systematically on every clear night. This is what the Rev. Robert Evans has done, and if you have, say, a 12-inch reflector, or even an 8-inch, you can certainly try your hand. You may never find a supernova; but you could quite conceivably find one tomorrow night.

There are some galaxies which are very powerful emitters of radio radiation, notably the Seyfert systems, which have bright, condensed nuclei and weak spiral arms (they are named after Carl Seyfert, who first drew attention to them as long ago as 1942). But as time went by, and radio astronomy 'came of age', it was found that some strong sources did not correspond with visible objects. In 1963 some of them were tracked down to what appeared to be faint bluish stars. When the optical spectra were examined, it was found that the objects were not stars at all; they were much more dramatic – super-luminous, very remote, and racing away from us at tremendous speeds. They were called Quasi-Stellar Objects or QSOs, though most people today refer to them as quasars.

Quasars, by no means all of which are radio sources, are now thought to be the nuclei of very active galaxies. The present holder of the distance record is a quasar identified in 1987 by astronomers using the telescopes at the Siding Spring Observatory in Australia; it is around 13 000 million light-years away, and is receding at well over 90 per cent of the velocity of light. Inevitably it is very faint, and the only quasar within range of average amateur telescopes is 3C–295, in Boötes (the Herdsman). It is rather below the 12th magnitude, and even when you find it it looks exactly like a star.

I have said that the quasars are receding from us. So are all the galaxies, apart from those contained in what we call the Local Group, a collection which includes the Andromeda Spiral, the Triangulum Spiral M.33 (on the fringe of naked-eye visibility), the Magellanic Clouds, a rather mysterious system called Maffei 1, and more than a couple of dozen dwarfs. (We know little about Maffei 1 because it lies near the main plane of our Galaxy, in Cassiopeia, and is heavily obscured by interstellar dust.) The entire universe is expanding, with each group of galaxies moving away from every other group, so that there is no well-defined centre. A particularly rich cluster of galaxies lies in the 'bowl' of Virgo, where you will find many systems of above the eleventh magnitude; their distances are of the order of 50 000 000 to 60 000 000 light-years.

The distances of the nearer galaxies can be found from their Cepheids. Further out, when we lose the Cepheids, we can make use of occasional supernovæ, since it is reasonable to assume that a supernova in an outer system will be about as luminous as a supernova in our Galaxy. Further out still, when we can no longer see individual stars or clusters, what is done is to measure the red shifts of the spectral lines, since –

according to what is termed Hubble's Law – the greater the shift, the greater the velocity of recession, and hence also the greater the distance. But there must be a limit, because we will eventually come to a distance at which a galaxy (or a quasar) is moving away at the full velocity of light: 186000 miles per second. In this case we will be unable to see it, and we will have reached the boundary of the observable universe, though not necessarily of the universe itself. So far as we can judge, this critical distance lies somewhere between 15000 million and 20000 million light-years, with a slight preference for the lower value. With the new Australian-found quasar, we must be approaching the limit.

It follows, too, that the Universe in its present form can be no more than 15000 million to 20000 million years old. Whether it began with a 'big bang', with all its material created at the same moment, or whether it is 'cyclic', alternately expanding and contracting, remains to be seen. Most astronomers today prefer the big bang idea, and certainly we can pick up weak radio emissions, coming in from all directions, which could represent the last effects of the original convulsion. But when we ask 'how' this happened, we have to admit that we do not have the slightest idea.

To say more about this sort of problem would lead me into theory – and I am not writing about theory; I am trying to show just what the amateur observer can see and do. Remember that astronomy today is not a science suited only to the professional researcher who has the advantages of a vast observatory together with ultra-modern electronics. You, too, can play a part, even if you have nothing more than a pair of binoculars or a portable telescope which you can set up in your back garden.

Appendix 1

Planetary Data

Planet	Distance from the Sun (millions of miles)			Sidereal period	Synodic period (days)	Axial rotation period (equatorial)		
	Maximum	Mean	Minimum					
Mercury	43	36	29	88 d	115.9	58 d	15 h	30 s
Venus	67.6	67.2	66.7	224.7 d	583.9	243.16 d		
Earth	94.6	93.0	91.4	365.3 d	—	23 h 56 min 4 s		
Mars	154.5	141.5	128.5	687 d	779.9	24 h 37 min 23 s		
Jupiter	506.8	483.3	459.8	11.86 y	398.9	9 h 50 min 30 s		
Saturn	937.6	886.1	834.6	29.46 y	378.1	10 h 39 min		
Uranus	1867	1783	1699	84.01 y	369.7	17 h 14 min		
Neptune	2817	2793	2769	164.79 y	367.5	17 h 52 min		
Pluto	4583	3666	2766	247.70 y	366.7	6 d 9 h 17 min		

Planet	Diameter miles, (equatorial)	Apparent Diameter (seconds of arc)		Maximum magnitude	Axial inclination	Escape velocity (miles/s)
		Max	.Min			
Mercury	3030	12.9	4.5	− 1.9	low	2.6
Venus	7523	66.0	9.6	− 4.4	178°	6.4
Earth	7926	—	—	—	23°27′	6.94
Mars	4218	25.7	3.5	− 2.8	23°59′	3.2
Jupiter	89424	50.1	30.4	− 2.6	3° 5′	37.1
Saturn	74914	20.9	15.0	− 0.3	26°44′	22.0
Uranus	31770	3.7	3.1	5.6	98°	13.9
Neptune	31410	2.2	2.0	7.7	28°48′	15.1
Pluto	1199	below 0.3		13.9	118°	very low

Appendix 2

Satellite Data

Satellite	Mean distance from centre of primary (thousands of miles)	Sidereal period			Orbital inclination (degrees)	Diameter (miles)	Maximum magnitude
		Days	Hours	Minutes			
Earth							
Moon	239.0	27	7	43	5.15	2160	−12.7
Mars							
Phobos	5.8		7	39	1.1	17 × 14 × 12	11.6
Deimos	14.6	1	6	18	1.8	10 × 7 × 6	12.8
Jupiter							
Metis	79.5		7	5	0	25	17.4
Adrastea	80.2		7	7	0	15 × 12 × 10	18.9
Amalthea	113		11	57	0.45	168 × 106 × 93	14.1
Thebe	138		16	12	0.9	68 × 62 × 56	15.5
Io	262	1	18	28	0.04	2263	5.0
Europa	417	3	13	14	0.47	1945	5.3
Ganymede	666	7	3	43	0.21	3274	4.6
Callisto	1170	16	16	32	0.51	2987	5.6
Leda	6895	239			26.1	9	20.2
Himalia	7135	251			27.6	106	14.8
Lysithea	7284	259			29.0	15	18.4
Elara	7295	260			24.8	50	16.7
Ananke	13176	631[a]			147	12	18.9
Carme	14046	692[a]			164	19	18.0
Pasiphaë	14605	735[a]			145	22	17.7
Sinope	14730	758[a]			153	17	18.3
Saturn							
Atlas	85.5		14	27	0.3	24 × 19 × 16	18.1
Prometheus	86.6		14	43	0.0	87 × 62 × 46	16.5
Pandora	88.1		15	6	0.1	68 × 52 × 41	16.3
Janus	94.1		16	41	0.1	118 × 121 × 56	14.5
Epimetheus	94.1		16	40	0.3	87 × 71 × 62	15.5
Mimas	115		22	37	1.52	247	12.9
Enceladus	148	1	8	53	0.02	310	11.8
Tethys	183	1	21	18	1.86	650	10.3

Satellite	Mean distance from centre of primary (thousands of miles)	Sidereal period			Orbital inclination (degrees)	Diameter (miles)	Maximum magnitude
		Days	Hours	Minutes			
Telesto	183	1	21	18	2	15 × 14 × 12	19.0
Calypso	183	1	21	18	2	19 × 15 × 10	18.5
Dione	235	2	17	41	0.02	696	10.4
Helene	235	2	17	41	0.2	22 × 21 × 17	18.5
Rhea	328	4	12	25	0.35	950	9.7
Titan	760	15	22	41	0.33	3201	8.4
Hyperion	920	21	6	38	0.43	218 × 145 × 124	14.2
Iapetus	2200	79	7	56	7.52	892	10
Phœbe	8050	550[a]	10	50	175	137	16.5
Uranus							
Cordelia	30.6		7	55	0	16	
Ophelia	33.1		8	55	0	19	
Bianca	36.7		10	23	0	26	
Cressida	38.4		11	7	0	39	
Desdemona	39.0		11	24	0	34	
Juliet	40.0		11	50	0	52	
Portia	41.1		12	19	0	67	
Rosalind	43.5		13	24	0	34	
Belinda	46.7		14	56	0	41	
Puck	53.4		18	17	0	89	
Miranda	81.1	1	9	50	4.22	293	16.5
Ariel	119	2	12	29	0.31	720	14.4
Umbriel	166	4	3	28	0.36	727	15.3
Titania	272	8	16	56	0.14	981	14.1
Oberon	364	13	11	7	0.10	946	14.2
Neptune							
N6	30.0	0	7	6	4.5	31	
N5	31.1	0	7	30	<1	56	
N3	32.6	0	8	0	<1	87	
N4	38.5	0	9	30	<1	99	
N2	45.7	0	13	18	<1	124	
N1	73.1	1	2	54	<1	260	
Triton	219.4	5[a]	21	3	159.9	1690	13.6
Nereid	3460	359	21	7	27.2	107	18.7
Pluto							
Charon	11.8	6	9	17	118	745	17.3

Note:

[a] Retrograde.

The satellites of Uranus revolve in the same sense as the planet itself − so that technically their motion is retrograde.

Appendix 3

The Brightest Asteroids

Asteroid	Mean distance from the Sun (thousands of miles)	Sidereal period (years)	Diameter (miles)	Mean opposition magnitude
1 Ceres	257.0	4.61	579	7.4
2 Pallas	257.4	4.61	325	8.0
3 Juno	247.8	4.36	168	8.7
4 Vesta	219.3	3.63	328	6.5
5 Astræa	239.3	4.14	73	9.8
6 Hebe	225.2	3.78	120	8.3
7 Iris	221.5	3.69	131	7.8
8 Flora	204.4	3.27	95	8.7
9 Metis	221.7	3.69	95	9.1
10 Hygeia	292.6	5.59	257	10.2
11 Parthenope	227.8	3.84	95	9.9
12 Victoria	217.2	3.56	80	9.9
14 Irene	240.4	4.16	98	9.8
15 Eunomia	246.0	4.30	154	8.5
16 Psyche	272.0	5.00	147	9.9
18 Melpomene	213.9	3.48	95	8.9
19 Fortuna	227.4	3.81	134	10.1
20 Massalia	224.1	3.74	80	9.2
23 Thalia	224.1	4.25	70	10.1
27 Euterpe	218.0	3.60	67	9.9
29 Amphitrite	236.7	4.08	160	10.0
42 Isis	226.9	3.81	60	10.2
43 Ariadne	204.6	3.27	50	10.2
44 Nysa	225.1	3.77	50	9.9
63 Ausonia	222.7	3.70	55	10.2

Note:
The only other asteroid which can reach magnitude 10.2 is 433 Eros, which on its rare close approaches can rise to 8.3 for a very brief period.

Appendix 4

Phenomena of Mercury, 1988–95

Inferior Conjunction

1988 Feb. 11, June 13, Oct. 11
1989 Jan. 25, May 23, Sept. 24
1990 Jan. 9, May 3, Sept. 8, Dec. 24
1991 Apr. 14, Aug. 21, Dec. 8
1992 Mar. 26, Aug. 2, Nov. 21
1993 Mar. 9, July 15, Nov. 6
1994 Feb. 20, June 25, Oct. 21
1995 Feb. 3, June 5, Oct. 5

Western Elongation

1988 Mar. 8, July 6, Oct. 26
1989 Feb. 18, June 18, Oct. 10
1990 Feb. 1, May 31, Sept. 24
1991 Jan. 14, May 12, Sept. 7, Dec. 27
1992 Apr. 23, Aug. 21, Dec. 9
1993 Apr. 5, Aug. 4, Nov. 22
1994 Mar. 19, July 17, Nov. 6
1995 Mar. 1, June 29, Oct. 20

Superior Conjunction

1988 Apr. 20, Aug. 3, Dec. 1
1989 Apr. 4, July 18, Nov. 10
1990 Mar. 19, July 2, Oct. 22
1991 Mar. 2, June 17, Oct. 3
1992 Feb. 12, May 31, Sept. 13
1993 Jan. 23, May 16, Aug. 29
1994 Jan. 3, Apr. 30, Aug. 13, Dec. 14
1995 Apr. 14, July 28, Nov. 23

Eastern Elongation

1988 Jan. 26, May 19, Sept. 15
1989 Jan. 9, May 1, Aug. 29, Dec. 23
1990 Apr. 13, Aug. 11, Dec. 6
1991 Mar. 27, July 25, Nov. 19
1992 Mar. 9, July 6, Oct. 31
1993 Feb. 21, June 17, Oct. 14
1994 Feb. 4, May 30, Sept. 26
1995 Jan. 19, May 12, Sept. 9

Note:
There will be a transit of Mercury on Nov. 6 1993. The next will be on 15 Nov. 1999.

Appendix 5

Phenomena of Venus, 1988–95

Inferior Conjunction

1988 June 12
1990 Jan. 18
1991 Aug. 22
1993 Apr. 1
1994 Nov. 2

Western Elongation

1988 Aug. 22
1990 Mar. 30
1991 Nov. 2
1993 June 10
1995 Jan. 13

Superior Conjunction

1989 Apr. 4
1990 Nov. 1
1992 June 13
1994 Jan. 17
1995 Aug. 20

Eastern Elongation

1988 Apr. 3
1989 Nov. 8
1991 June 13
1993 Jan. 19
1994 Aug. 24

Note:
No transits of Venus occur during the twentieth century. The next will be on June 7 2004 and June 4 2012, after which there will be no more until Dec. 10 2117 and Dec. 8 2125.

Appendix 6

Oppositions of the Superior Planets, 1988–95

Date	Constellation	Maximum magnitude	Maximum apparent diameter, (seconds of arc)	Opposition distance from Earth (millions of miles)
Mars				
1988 Sept. 22	Pisces	− 2.8	23.8	35.3
1990 Nov. 27	Taurus	− 1.8	18.1	47.8
1993 Jan. 7	Gemini	− 1.2	14.9	58.4
1995 Feb. 12	Cancer	− 1.0	13.9	62.8
Jupiter				
1988 Nov. 23	Taurus	− 2.4	48.8	375
1989 Dec. 27	Gemini	− 2.3	47.3	387
1991 Jan. 29	Cancer	− 2.1	45.7	400
1992 Feb. 29	Leo	− 2.0	44.6	410
1993 Mar. 30	Virgo	− 2.0	44.2	414
1994 Aug. 30	Libra	− 2.0	44.5	411
1995 June 1	Scorpius	− 2.1	45.5	402
Saturn				
1988 June 20	Sagittarius	+ 0.2	18.5	840
1989 July 2	Sagittarius	+ 0.2	18.5	839
1990 July 14	Sagittarius	+ 0.3	18.5	836
1991 July 27	Capricornus	+ 0.3	18.6	832
1992 Aug. 7	Capricornus	+ 0.4	18.8	826
1993 Aug. 19	Capricornus	+ 0.5	18.9	818
1994 Sept. 1	Aquarius	+ 0.7	19.1	809
1995 Sept. 14	Aquarius	+ 0.9	19.4	800
Uranus				
1988 June 20	Ophiuchus/Sagittarius	+ 5.9	3.7	1710
1989 June 24				
1990 June 29				

Date	Constellation	Maximum magnitude	Maximum apparent diameter, (seconds of arc)	Opposition distance from Earth (millions of miles)
1991 July 4				
1992 July 7				
1993 July 12				
1994 July 17				
1995 July 21				
Neptune				
1988 June 30	Sagittarius	+ 7.7	2.5	2711
1989 July 2				
1990 July 5				
1991 July 8				
1992 July 9				
1993 July 12				
1994 July 14				
1995 July 17				
Pluto				
1988 May 1	Virgo	+ 14	below 0.3	2666
1989 May 4				
1990 May 7				
1991 May 10				
1992 May 12				
1993 May 14				
1994 May 17				
1995 May 20				

THE ANTONIADI SCALE OF SEEING FOR PLANETARY WORK

1.   Perfect seeing, without a quiver.
2.   Slight undulations, with moments of calm lasting for several seconds.
3.   Moderate seeing, with larger air tremors.
4.   Poor seeing, with constant troublesome undulations.
5.   Very bad seeing, unsuitable for anything except, possibly, a very rough sketch.

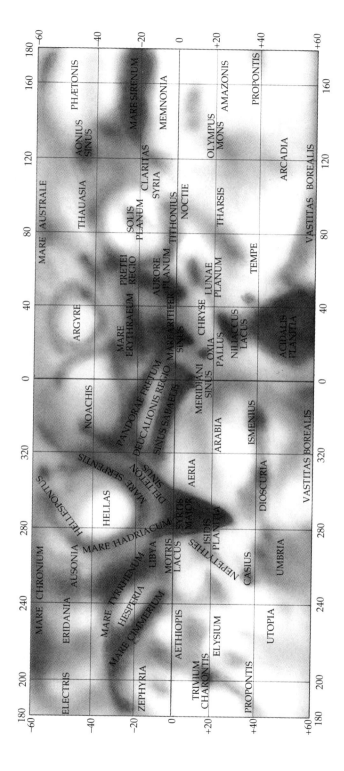

Fig. A6.1. Map of Mars, drawn from my observations made between 1970 and 1989 with my 15-inch and 12½-reflectors.

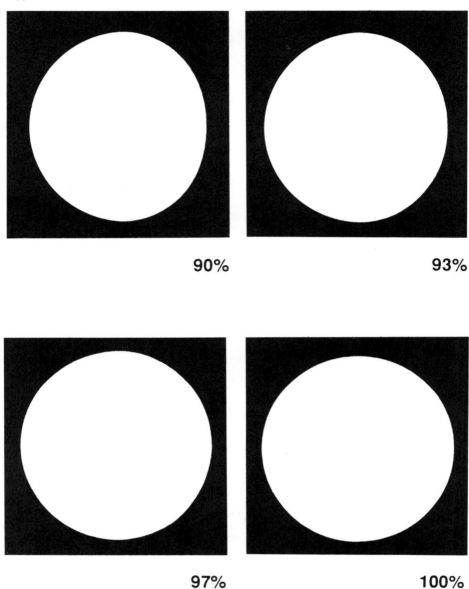

Fig. A6.2. *'Blanks' for Mars observers. These can be photocopied and used at the telescope. Except when Mars is near opposition, the phase is always appreciable, and must be allowed for when an accurate sketch is to be made.*

Appendix 7

Jupiter: Transit Observations

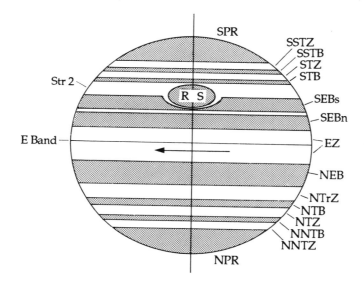

Fig. A7.1. The main belts and zones of Jupiter: SPR = South Polar Region; SSTZ = South South Temperate Zone; SSTB = South South Temperate Belt; STZ = South Temperate Zone; STB = South Temperate Belt; STrZ = South Tropical Zone; SEB = South Equatorial Belt (often double: SEBs = south component, SEBn = north component); EZ = Equatorial Zone; EB = Equatorial Band; NEB = North Equatorial Belt (often double: NEBs, NEBn); NTrZ = North Tropical Zone; NTB = North Temperate Belt; NTZ = North Temperate Zone; NNTB = North North Temperate Belt; NNTZ = North North Temperate Zone; NPR = North Polar Region.

The following is a typical extract from my observation diary:

1963 November 4, 12½in. refl. Seeing 2 to 3

		Longitude		
GMT	Feature	System I	System II	Comments
19.57	c. of white spot in STZ	. . .	182.1	× 360
20.06	f. of this white spot	. . .	187.5	
20.21	c. of white patch in middle of EqZ	253.4	. . .	
20.22	p. of visible section of NTB	. . .	197.2	
20.37	f. of dark mass on N. edge of NEB	263.1	. . .	

(c = centre, p = preceding, f = following)

To work out the longitudes of the features, use the tables given in a yearly almanac, which given the longitude of the central meridian for 0h for every hour of every day.

Example. The 19.57 transit of the centre of the white spot in the STZ. From the tables: longitude of the central meridian (System II) for 16 h on November 4 is 038.9. This is 3 h 57 m earlier than the time of the transit. Therefore, the longitude for 19.57 may be worked out from the tables given below:

Longitude at	16 h	Nov. 4:	= 038.9
	+3 h		= 108.8
	+	50 min	= 030.2
	+	7 min	= 004.2
	+3 h 57 min		182.1

If the calculated longitude works out at over 360, then subtract 360°.

It is important to use the correct System: System I is bounded by the north edge of the SEB and the south edge of the NEB, all the rest of the planet being System II. If the wrong System is used, the results can be very peculiar indeed!

Change of longitude in intervals of mean time

Time	System I	System II	Time	System I	System II
1 min	0.6	0.6	1 hour	36.6	36.3
2	1.2	1.2	2	73.2	72.5
3	1.8	1.8	3	109.7	108.8
4	2.4	2.4	4	146.3	145.1
5	3.0	3.0	5	182.9	181.3
6	3.7	3.6	6	219.5	217.6
7	4.3	4.2	7	256.1	253.8
8	4.9	4.8	8	292.7	290.1
9	5.5	5.4	9	329.2	326.4
10	6.1	6.0	10	365.8	362.6
20	12.2	12.1			
30	18.3	18.1			
40	24.4	24.2			
50	30.5	30.2			

Fig. A7.2. Blank for drawing Jupiter at the telescope.

Appendix 8

Saturn: Intensity Estimates

Prominent spots on Saturn are rare. However, useful work can be done in estimating the brightness of the different features, as these are certainly variable. The scale adopted is from 0 (brilliant white) to 10 (black shadow). In general, Ring B is the brightest feature, and the outer part has a value of 1.

The easiest way of recording is to prepare a sketch (perhaps a rough one) and then merely jot down the intensities on the drawing itself. It is best to make each estimate twice: first, start with the darkest feature and work through to the brightest; then begin once more, this time with the brightest feature. The following is an extract from my own notebook:

1987 August 15, 20.30 to 20.50 GMT. 15in. refl. × 360. Seeing 3.

Ring B, outer	1
Ring B, inner	2
Equatorial Zone	2
N. Temperate Zone	3
N. Polar Region	3.5 ·
N. Temperate Belt	4
Ring A	5
N. Equatorial Belt	6; region between components, 5
Encke Division	6.5
Ring C	7
Cassini Division	7.5
Shadow, Ring on Globe	9
Shadow, Globe on Ring	10

The S. polar region could not be seen, as it was covered by the rings.

During the mid-1980s the rings are wide open, and the northern hemisphere of the planet is uncovered. The rings will narrow steadily, and become edgewise-on once more in 1995–6. The Earth will pass through the ring plane on May 21 and Aug. 11 1995 and Feb. 11 1996; the Sun will pass through the ring plane Nov. 19 1995.

Fig. A8.1. Blanks for drawing Saturn at the telescope.

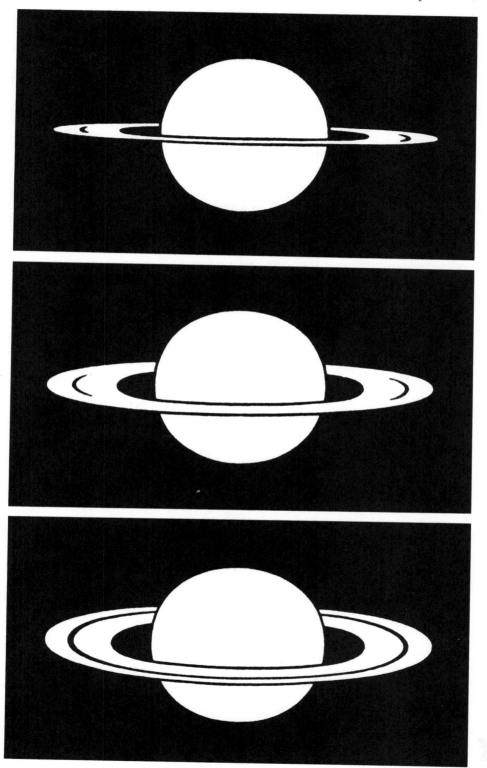

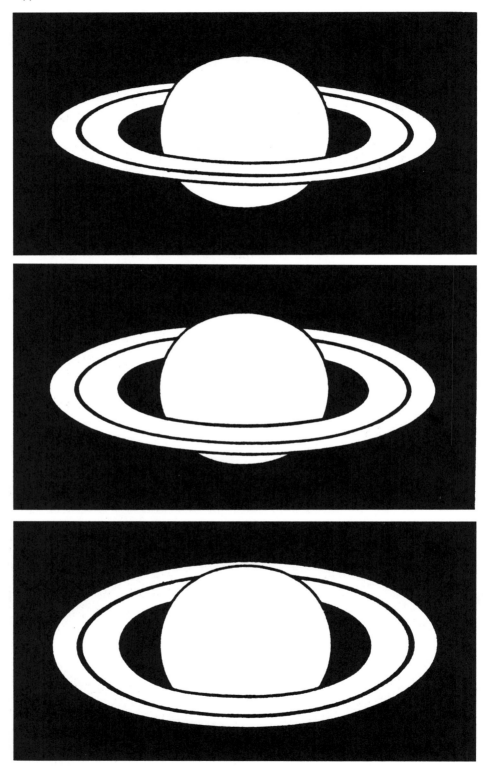

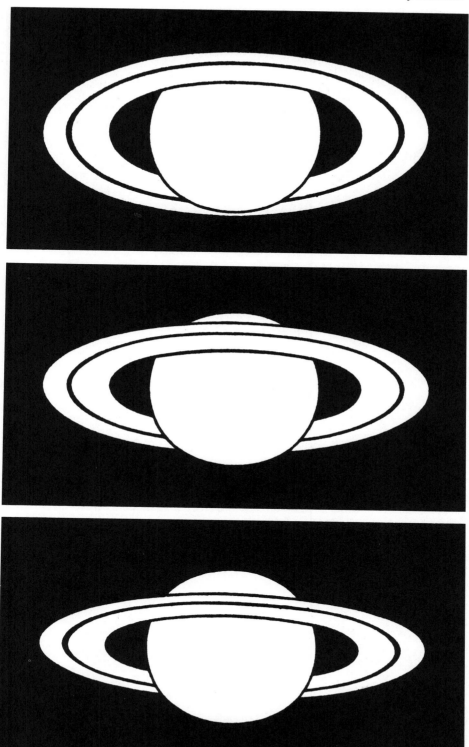

Appendix 9

Observing the Sun

As I have said in the text, I restrict myself entirely to the method of projection. As I live in the Earth's northern hemisphere, the projected image has N. at the top, S. at the bottom, E. to the right and W. to the left. If the telescope is not being clock-driven, the Sun's image will drift to the left, and spots will appear at the right limb and be carried from right to left by the Sun's rotation, finally disappearing over the left limb.

A scale of 6 in to the Sun's diameter is recommended. Orient it as described, by allowing a spot to drift along a diameter line which has been drawn on your circle. You then need three co-ordinates:

P = the position angle of the N. end of the Sun's axis of rotation; + if east, and − if west.

B_0 = the latitude of the centre of the Sun's disk, + or −, indicating the tilt of the axis toward or away from the Earth.

L_0 = the longitude of the centre of the disk, or of the central meridian.

P and B_0 change slowly; the value for a standard time each day can be looked up from a yearly almanac. L_0, however changes quickly by a rate of 13.2 degrees per day. The value for noon each day is given in the almanacs, and an adjustment must be made to fit in with the time of your observation. The table given here is only rough, but will help as a guide:

Decrease of L_0 with time

Time	Decrease (degrees)	Time	Decrease (degrees)
5 min	0.1	10	5.5
30 min	0.3	12	6.6
1 h	0.6	14	7.7
2	1.1	16	8.8
3	1.7	18	9.9
4	2.2	20	11.0
5	2.7	22	12.1
6	3.3	24	13.2
8	4.4		

Appendix 10

Eclipses of the Sun, 1988–95

Date	GMT	Type	Maximum duration of totality	Area of visibility
1988 Mar. 18	02	Total	3 m 46 s	Pacific, East Indies, Borneo
1989 Mar. 7	18	Partial, 83%	—	Arctic
1989 Aug. 31	06	Partial, 63%	—	Antarctic
1990 Jan. 26	20	Annular	—	Antarctic, South Atlantic
1990 July 22	03	Total	2 m 33s	Finland, Russia, Pacific
1991 Jan. 15	24	Annular	—	New Zealand, Pacific, SW Australia
1991 July 11	19	Total	6 m 54 s	Hawaii, Mexico, Pacific, Brazil
1992 Jan. 4	23	Annual	—	Central Pacific
1992 June 30	12	Total	5 m 20 s	South Atlantic
1992 Dec. 24	01	Partial, 84%	—	Arctic
1993 May 21	14	Partial, 74%	—	Arctic
1993 Nov. 13	22	Partial, 93%	—	Antarctic
1994 May 10	17	Annular	—	N. America, Atlantic, Eastern Pacific, extreme NW Africa
1994 Nov. 3	14	Total	4 m 24 s	Peru, Brazil, S. Atlantic
1995 Apr. 29	18	Annular	—	Peru, S. Atlantic, S. Pacific
1995 Oct. 24	05	Total	2 m 10s	Iran, India, East Indies, Pacific

Appendix 11

Eclipses of the Moon, 1988–95

Date	GMT of mid-eclipse	Phase	Maximum duration of totality
1988 Mar. 3	16:14	Penumbral	—
1988 Aug. 27	11:05	Partial, 29%	—
1989 Feb. 20	15:37	Total	1 h 16 min
1989 Aug. 17	03:09	Total	1 h 38 min
1990 Feb. 9	19:13	Total	0 h 46 min
1990 Aug. 6	14:12	Partial, 68%	
1991 Jan. 30	06:00	Penumbral	—
1991 June 27	03:16	Penumbral	—
1991 July 26	18:09	Penumbral	—
1991 Dec. 21	10:34	Partial, 9%	—
1992 June 15	04:58	Partial, 68%	—
1992 Dec. 9	23:45	Total	1 h 14 min
1993 June 4	13:02	Total	1 h 38 min
1993 Nov. 29	06:26	Total	0 h 50 min
1994 May 25	03:32	Partial, 24%	—
1994 Nov. 18	06:45	Penumbral	—
1995 Apr. 15	12:19	Partial, 11%	—
1995 Oct. 8	16:05	Penumbral	—

Appendix 12

The Limiting Lunar Detail Visible with Various Apertures

The following table (based upon work by E.A. Whitaker) gives the approximate diameters of the smallest crater half-filled with shadow, and of the narrowest black line certainly distinguishable. Perfect seeing conditions and optical equipment are assumed. Obviously, different observers will have different limits; but using the Meudon 33-inch refractor (Paris Observatory) during the pre-Apollo period, when I was engaged in Moon-mapping, I recorded a craterlet on the summit of a mountain peak near Beer that was only about 500 yards in diameter.

Aperture of OG (inches)	Smallest crater (miles)	Narrowest rill
1	9	0.5 miles
2	4.5	0.25 miles
3	3	0.16 miles
4	2.25	220 yards
6	1.5	150 yards
8	1.1	110 yards
10	0.9	90 yards
12	0.75	70 yards
15	0.6	60 yards
18	0.5	50 yards
33	500 yards	30 yards

Appendix 13

The Lunar Maps

Finding one's way around the surface of the Moon needs considerable patience, because the aspects of the various features alter so quickly with the changing angle of solar illumination, but the photographs and maps given here should be good enough to enable you to recognize the main craters and mountains, plus the broad grey 'seas'. In the photographs of the two hemispheres, the formations near the eastern and western limbs are under high light, and are not well seen; for example Petavius, in the south-east, is really a majestic crater 100 miles across, but under this lighting it is hard to make out at all. Changing libration also has marked effects upon formations which are anywhere near the limb.

The notes given here are necessarily brief, and I have limited them to fairly obvious 'landmarks' plus some formations of special interest, such as Linné, the Alpine Valley and the Straight Wall.

NAMES OF LUNAR SEAS (MARIA)

Sinus Æstuum:	Bay of Heats	Mare Nectaris:	Sea of Nectar
Mare Australe:	Southern Sea	Mare Nubium:	Sea of Clouds
Mare Crisium:	Sea of Crises	Mare Orientale:	Eastern Sea
Palus Epidemiarum:	Marsh of Epidemics	Oceanus Procellarum:	Ocean of Storms
Mare Fœcunditatis:	Sea of Fertility	Palus Putredinis:	Marsh of Decay
Mare Frigoris:	Sea of Cold	Sinus Roris:	Bay of Dews
Mare Humboldtianum:	Humboldt's Sea	Mare Serenitatis:	Sea of Serenity
Mare Humorum:	Sea of Humours	Mare Smythii:	Smyth's Sea
Mare Imbrium:	Sea of Showers	Palus Somnii:	Marsh of Sleep
Sinus Iridum:	Bay of Rainbows	Lacus Somniorum:	Lake of the Dreamers
Mare Marginis:	Marginal Sea	Mare Spumans:	Foaming Sea
Sinus Medii:	Central Bay	Mare Tranquillitatis:	Sea of Tranquillity
Lacus Mortis:	Lake of Death	Mare Undarum:	Sea of Waves
Palus Nebularum:	Marsh of Mists	Mare Vaporum:	Sea of Vapours

THE EASTERN HALF OF THE MOON

(1) North-eastern quadrant

This quadrant contains three major seas, Serenitatis, Tranquillitatis and Crisium, together with parts of the Mare Vaporum and the Mare Frigoris, as well as the northern part of the Mare Fœcunditatis. Mare Serenitatis is the most conspicuous: it contains various craterlets, of which the most prominent is Bessel; there is also the famous (or notorious) Linné, and a number of ridges. Near full moon, the Mare is seen to be crossed by a bright ray which passes by Bessel. Mare Tranquillitatis is lighter and patchier; between it and Serenitatis there is a strait on which lies the magnificent 30-mile crater Plinius, which has two interior craterlets near its centre. The Mare Crisium is well-defined; note the curious colour of the Palus Somnii, which is bordered by the brilliant Proclus. Close to the eastern limb are two very foreshortened seas, Mare Smythii and Mare Marginis, while Mare Humboldtianum lies further north; these can be well seen only during conditions of favourable libration.

Of the mountain ranges, the most important are those bordering the Mare Serenitatis: the Hæmus and the Caucasus Mountains, with peaks rising to 8000 feet and 12000 feet, respectively. Part of the Alps can be seen, cut through by the Alpine Valley – a most interesting feature, not like anything else on the Moon. South of the Hæmus, and north of the large ruined plain Hipparchus, can be seen the two most important rills or clefts of the Mare Vaporum – those of Hyginus (which is basically a crater-chain) and Ariadæus. Both these can be seen with a very small telescope when suitably illuminated. The major lunar craters are listed below:

Arago. Diameter 18 miles. It lies on the Mare Tranquillitatis, and has a low central elevation. Near it are several domes.

Archytas. A bright 21-mile crater on the north edge of the Mare Frigoris. It has a central peak.

Aristillus. Diameter 35 miles, with walls rising to 11000 feet above the floor. The walls are bright, and there is a central peak. Aristillus forms a notable pair with Autolycus.

Aristoteles. A prominent 52-mile crater, forming a pair with Eudoxus.

Atlas. Forms a pair with Hercules; it lies north of the Mare Serenitatis. Diameter 55 miles. The walls are terraced, rising in places to 11000 feet. There is considerable detail on the floor.

Autolycus. The companion to Aristillus. Autolycus is 24 miles in diameter, and 9000 feet deep.

Boscovich. On the Mare Vaporum. It is a low-walled, irregular formation, recognizable (like its companion Julius Cæsar) by its very dark floor.

Bürg. A 28-mile crater between Atlas and Aristoteles, with a large central peak on which is a summit craterlet. West of Bürg lies an old plain crossed by an elaborate system of rills.

Cassini. On the fringe of the Alps. A curious broken formation, shallow, and 36 miles in diameter. It contains a prominent craterlet, Cassini A.

Cleomedes. A 78-mile crater near the Mare Crisium. It is broken in the west by a smaller but deeper crater, Tralles.

Dionysius. A brilliant small crater near Sabine and Ritter.

Endymion. This 78-mile crater can always be recognized because of its dark floor. Patches inside it are somewhat variable.

Eudoxus. The companion of Aristoteles. It is 40 miles in diameter, and 11 000 feet deep.

Firminicus. Closely south of the Mare Crisium. It is dark-floored with a diameter of 35 miles.

Gauss. A magnificent 100-mile crater, very conspicuous when on the terminator - though it is, of course, highly foreshortened..

Geminus. Diameter 55 miles, near Cleomedes. Its lofty walls are strongly terraced.

Godin. Diameter 27 miles, near Ariadæus and Hyginus. Closely north of it is Agrippa, which is slightly larger but rather less deep.

Hercules. The companion of Atlas. It is 45 miles in diameter, with terraced walls which appear brilliant times. Inside it is a larger craterlet, Hercules A.

Julius Cæsar. A low-walled formation in the Mare Vaporum area. Like Boscovich, it has a dark floor which makes it identifiable at any time.

Macrobius. A 42-mile crater near the Mare Crisium, with walls rising to 13 000 feet. There is a low, compound central mountain mass.

Manilius. A 25-mile crater on the Mare Vaporum, notable because of its brilliant walls.

Menelaus. Another brilliant crater; it is 20 miles in diameter, lying in the Hæmus Mountains. Like Manilius, it is so bright that it is easy to identify.

Posidonius. A 62-mile plain on the border of the Mare Serenitatis. Adjoining it to the east is a smaller, squarish formation, Chacornac, and south is Le Monnier, one of the 'bays' with a broken-down seaward wall. The Russian probe Luna 21 landed near here, carrying the 'crawler' Lunokhod 2 which operated from January to May 1973 – and is still there, though now inactive.

Proclus. Closely west of the Mare Crisium. This is one of the most brilliant formations on the Moon, and is the centre of a ray system which is asymmetrical; two of the rays mark the border of the Palus Somnii.

Sabine and Ritter. Two 18-mile craters on the west border of the Mare Tranquillitatis. The first lunar landing, with Apollo 11, was made some distance east of the pair.

Scoresby. A very distinct crater 36 miles in diameter, near the North Pole. It is much the most conspicuous formation in this area, and is very useful as a landmark..

Taruntius. A 38-mile crater south of the Mare Crisium, with narrow walls and a low central hill. It is a typical 'concentric crater', with a complete inner ring.

(2) South-eastern quadrant

This quadrant is occupied largely by rugged uplands, and large and small craters abound. The only major seas are the small, well-marked Mare Nectaris and most of the larger Mare Fœcunditatis; the Mare Australe, near the limb, is much less well-defined. The only mountains of note are near the limb. The so-called Altai Mountains really form a scarp rather than a range, and are associated with the Mare Nectaris system.

Albategnius. A magnificent walled plain near the centre of the disk; the companion of Hippachus. Diameter 80 miles. The south-west wall is disturbed by a deep 20-mile crater, Klein.

Capella. A 30-mile crater near Theophilus. It has a very massive central mountain, topped by a summit craterlet; the floor of Capella is crossed by a deep valley. It has a shallower companion, Isidorus.

Cuvier. This forms an interesting group with Licetus and the irregular Heraclitus. Cuvier is 50 miles across.

Fabricius. A 55-mile crater, similar in size to its companion, Metius. Fabricius interrupts the vast ruined plain, Janssen.

Gutenberg. This and its companion Goclenius lie on the high region between Mare Nectaris and Mare Fœcunditatis. To the north lie some delicate rills.

Hipparchus. Hipparchus is 84 miles in diameter, and adjoins Albategnius; it is very conspicuous when shadow-filled, but its walls are low and broken, so that when away from the terminator it is not easy to identify. Ptolemæus lies closely west of it.

Langrenus. An 85-mile crater, with high walls and central mountain. It is a member of the great Eastern Chain, which extend from Furnerius in the south and includes Petavius, Vendelinus, Mare Crisium, Cleomedes, Geminus and Endymion.

Maurolycus. A deep 75-mile crater in the far south. West of it lies the larger Stöfler, which has a darkish floor.

Messier. This curious little crater on the Mare Fœcunditatis is one of a pair of nonidentical twins; it companion, Messier A, is rather larger and less circular. Extending to the west is a strange ray rather like a comet's tail.

Petavius. A magnificent crater, with a central mountain group and a prominent rill running across the floor. Closely east of Petavius is Palitzsch, often described as a valley-like structure, but which is in fact a crater-chain.

Piccolomini. A 56-mile crater south of Fracastorius and the Mare Nectaris. It is deep and conspicuous; it lies at the eastern end of the Altai Scarp.

Rheita. A 42-mile crater. Associated with it is the famous Rheita Valley, one of the most important crater-chains on the Moon; there is another formation of the same sort not far off, associated with the crater Reichenbach.

Steinheil. This 42-mile crater forms a pair of 'Siamese twins' with its similar but shallower neighbour, Watt. They lie not far from the Mare Australe.

Theophilus. The northern member of the grand chain of which Cyrillus and Catharina are the other members. Theophilus is 65 miles across and 18000 feet deep; it is one of the most magnificent of all the lunar craters, with massively terraced walls and a lofty, complex central mountain group.

Vendelinus. One of the Eastern Chain. It is 100 miles across, but is comparatively low-walled, and is conspicuous only when near the terminator.

Vlacq. One of a group of large ring-plains near Janssen; 56 miles in diameter, and 10000 feet deep.

Werner. A 45-mile crater. It forms a pair with its neighbour Aliccensis, and nearby are three more pairs; Apian–Playfair, Azophi–Abenezra, and Abulfeda–Almanon..

Wilhelm Humboldt. On the eastern limb, near Petavius. It is 120 miles across, but very foreshortened. Space-probe pictures show a mass of detail inside it, quite invisible from Earth

THE WESTERN HALF OF THE MOON

(3) North-west quadrant

This quadrant consists largely of 'sea'; there is the Mare Imbrium, over 700 miles across, as well as parts of the even vaster but less well-defined Oceanus Procellarum, plus most of the Mare Frigoris and the Sinus Roris. The chief mountains are the Apennines – much the most conspicuous range on the Moon – the Alps and the Carpathians, making up another part of the Imbrian border. Near full moon, the most striking features are the brilliant Copernicus, Aristarchus and Kepler, and the dark-floored Plato. The Alpine Valley is of special interest.

Anaxagoras. A 32-mile crater not far from the North Pole. It is the centre of a ray-system, and is always distinct.

Archimedes. A 50-mile plain on the Mare Imbrium, with low but regular walls. It forms a splendid trio with Aristillus and Autolycus.

Aristarchus. The brightest formation on the Moon. Nearby is its darker-floored companion Herodotus, together with the great winding valley known as Schröter's Valley; to the north-east are the discontinuous clumps of the Harbinger Mountains. More TLP have been reported in the Aristarchus area than anywhere else, so that it should be closely watched.

Copernicus. The great 56-mile ray-crater, whose system is inferior only to Tycho's. With its massive walls and complex central mountain group, Copernicus well earns its nickname of 'the Monarch of the Moon'.

Eratosthenes. A deep, 38-mile crater at the end of the Apennines, bordering Sinus Æstuum. It has lofty, terraced walls, and the patches on the floor seem to be variable.

Sinus Iridum. The famous 'bay' leading out of the Mare Imbrium. The old seaward wall is now marked only by low, discontinuous ridges. The capes to either side are known as Heraclides and Laplace; the mountains toward the limb − the Jura Mountains − catch the Sun's rays before the bay itself, producing the Jewelled Handle effect at sunrise over the area.

Kepler. A 22-mile crater on the Oceanus Procellarum; the centre of a major ray-system. South of it is a crater of much the same size, Encke, which is shallower and is not a ray-centre.

Olbers. Another ray-centre, near the limb north of Grimaldi. Although it is so foreshortened, it can be very conspicuous.

Philolaus. A crater 46 miles across, near the limb, forming a pair with its neighbour Anaximenes. Reddish hues have been reported inside Philolaus, probably indicating some unusual surface material.

Pico. An 8000-foot mountain on the Mare Imbrium, south of Plato, with at least three peaks. Some way to the south-east is Piton, with a summit craterlet.

Plato. This regular, 60-mile formation has a dark floor, and is one of the most-studied craters on the Moon; inside it are some delicate craterlets which show baffling changes in visibility, and the hue of the floor itself is variable. Plato is always identifiable, and will repay close, continuous attention.

Pythagoras. A very deep crater 85 miles in diameter, badly foreshortened, but magnificent when best seen. There are numerous large formations near the limb in this area.

Stadius. A 'ghost' crater very close beside Eratosthenes. The two are about the same size, but the walls of Stadius have been so reduced that they are now barely traceable even under ideal conditions of illumination.

Straight Range. A peculiar line of peaks on the Mare Imbrium, near Plato. It is 40 miles long, and rises to 6000 feet.

Timocharis. A 23-mile crater on the Mare Imbrium; it has a central craterlet, and is the centre of an inconspicuous system of rays.

(4) South-west quadrant

This quadrant is crammed with interesting features. In the northern part of it lie the well-marked Mare Humorum, part of the Oceanus Procellarum, and most of the vast Mare Nubium; the southern part is mainly rough upland. The Mare Orientale is so close to the limb that it is never easy to identify, though we now know it to be a huge ringed structure extending onto the far side of the Moon. The quadrant includes some mountains – the Riphæans, on the Mare Nubium, and the Percy Mountains, forming part of the border of the Mare Humorum.

Alphonsus. The great walled plain close to Ptolemæus. Dark patches lie on its floor; there is a low central peak, and a system of rills. The first spectrographically-confirmed TLP was seen in Alphonsus, in 1958, by N.A. Kozyrev. It is adjoined to the south by the smaller, deeper Arzachel.

Bailly. Aptly described as a 'field of ruins', almost 180 miles across, but right on the limb, so that it is never well seen.

Billy. A 30-mile crater south of Grimaldi. It can always be identified because of its dark floor; its near neighbour, Hansteen, is similar in size but much less dark.

Bullialdus. A splendid 32-mile crater on the Mare Nubium, with terraced walls and a central peak. This is one of the most perfect of all the ring-plains.

Clavius. Clavius is 145 miles across, and when right on the terminator can be identified with the naked eye. The walls contain peaks rising to 17 000 feet above the deepest part of the floor. Inside Clavius is a chain of craters, decreasing in size from east to west.

Crüger. A low-walled, regular crater 30 miles in diameter, near Grimaldi. Like Billy, it is easy to find because of its dark floor.

Doppelmayer. An interesting 40-mile bay on the Mare Humorum. The seaward wall can just be traced, and there is a very reduced central mountain.

Euclides. Only 7 miles in diameter, but surrounded by a prominent bright nimbus. It lies near the low, irregular Riphæan Mountains.

Fra Mauro. One of a group of damaged ring-plains on the Mare Nubium; the others in the group are Parry, Bonpland and Guericke. Apollo 14 landed in this region.

Gassendi. A magnificent walled plain on the northern border of the Mare Humorum. It is 55 miles in diameter, with a central mountain group; there is considerable floor detail, including a system of rills. Gassendi is one of the best-known TLP sites.

Hippalus. Another bay on the Mare Humorum, not unlike Doppelmayer. Its floor is crossed by rills, and there is a rill-system nearby which is easily visible with a small telescope.

Grimaldi. Identifiable at all times because of its dark floor – generally, in fact, the darkest area on the whole of the Moon. Grimaldi is 120 miles across, with low walls. Adjoining it is Riccioli, 80 miles across, inside which is a very dark patch.

Letronne. A bay 70 miles in diameter, lying on the edge of the Oceanus Procellarum not far from Grimaldi. There is the remnant of a central elevation.

Maginus. A very large walled plain near Clavius and Tycho. It is prominent when near the terminator, but very obscure under high light.

Mercator. This and Campanus form a conspicuous pair of craters east of the Mare Humorum. Each is about 28 miles in diameter, and the only obvious difference between them is that Mercator has the darker floor.

Mersenius. A convex-floored 45-mile crater near Gassendi, asssociated with an interesting system of rills.

Moretus. A splendid crater 75 miles in diameter and 15 000 feet deep; it is a pity that it lies so near the limb. It has an exceptionally high central mountain.

Pitatus. Pitatus has been described as being 'like a lagoon'. It lies on the southern border of the Mare Nubium, and has a dark floor containing a low central mountain. It is 50 miles in diameter. West of it is a smaller formation, Hesiodus, from which a very prominent rill runs toward Mercator and Campanus.

Ptolemæus. Over 90 miles across, and one of the most celebrated formations on the Moon, if only because it is nearly central on the disk. It has a darkish, relatively smooth floor, containing only one conspicuous crater. Ptolemæus is the northern member of a chain of three great walled plains, the others being Alphonsus and Arzachel. South of this chain lies another, made up of Purbach, Regiomontanus and Walter.

Schickard. A walled plain 134 miles in diameter, with fairly regular walls and darkish interior patches. It is the largest member of a group which includes Phocylides, Nasmyth and Wargentin.

Sirsalis. This and its 'Siamese twin' lie near the dark-floored Crüger, and are associated with one of the most important of all the lunar rills.

Straight Wall. The celebrated fault in the Mare Nubium, near Birt. It lies inside a large, very obscure ring.

Thebit. A 37-mile crater in the Straight Wall area. It is interrupted by a smaller crater, which is in turn interrupted by a third. This group is a useful test object for telescopes of small aperture.

Tycho. The major ray-centre of the Moon – a 54-mile crater, with massive, terraced walls and complex central structure. Near full, the Tycho rays dominate the whole scene.

Vitello. A 30-mile crater on the edge of the Mare Humorum, with an inner but not quite concentric ring.

Wargentin. Near Schickard; it is 55 miles in diameter, and is much the finest example of a lunar plateau. It is lava-filled, so that its 'floor' is raised. It is most unfortunate that Wargentin lies so near the limb.

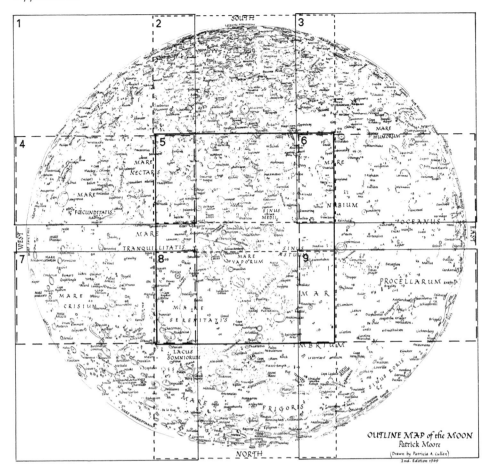

OUTLINE MAP of the MOON
Patrick Moore
(Drawn by Patricia A. Cullen)
2nd. Edition 1969

SOUTH

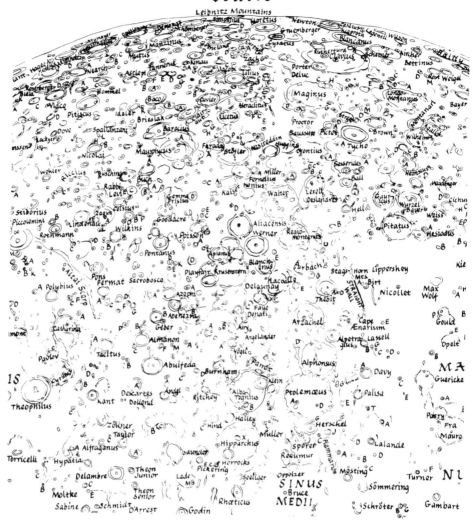

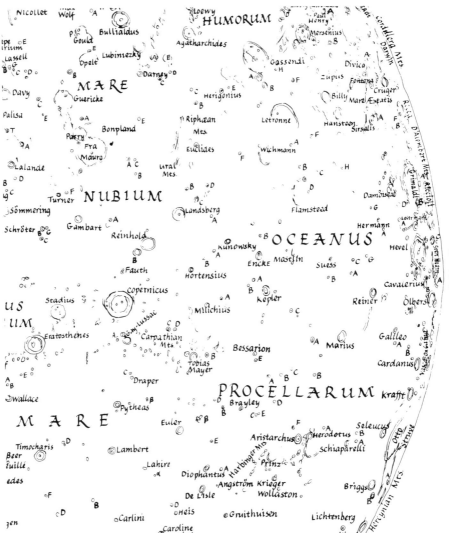

EAST

MARE
Schubert UNDARUM
Apollonius
Firmicus
Azout
Da Vinci
Cauchy
Sinas
E
B
Sosigenes
Macli
Ross
D
Condorcet
Barker's
Quadrangle
Lick
Glaisher
Yerkes
Lyell
F
Jansen
Plinius
A
Hæmu
Mts.
Ta
Neper
Picard
Cape
Lavinium
Palus
Somnii
A
Franz
A
Vitruvius
L
Dawes
C
Cape
Argæus
Cape
Olivium
Proclus
E
D
Achurusia
Hansen Agarum
Cape
MARE
MARGINIS
D
Peirce
Maraldi
Deseilligny
Be
Alhazen
M A R E
A
C
Macrobius
B
Littrow
B
M A R
Tisserand
E
J
K
MARE
ANGUIS
D
S E R E N
Oriani
Plutarch
G
Eimmart
Romer
Le Monnier
Chacornac
Posidonius
Delmotte
Cleomedes
B
Tralles
H
A
Taurus
Mts.
A
P
Luther
Oseneca
Debes
G
D
Newcomb
Kirchhoff
B
J
Daniell
Hahn
Burckhardt
Geminus
Taurus
Mts.
G.P.Bond
Hall
B
C
K
LACUS
Timoleon
Berosus
Bernoulli
Berzellius
A
F
Maury
Franklin
SOMNIORUM
Grove
Gauss
F
Messala
A
B
Cepheus
E
Plana
D
Hooke
K
J
Mason
Burg
La
Mo
Struve
Schumacher Schuckberg
Carrington
Ørsted
A
Hercules
A
Zeno
D
Chevallier
Atlas
Williams
Baily
Mercurius
F
H
J
A
Endymion
D
A
M A
A
De la Rue
A
Gärtr
MARE HUMBOLDTIANUM
A
Strabo Thales
T
N B
Schwabe
A Cusa
P

NORTH

Appendix 14

Some of the more important Periodical Comets

| Comet | Period (y) | Distance from Sun (au) | |
		Perihelion	Aphelion
Encke	3.3	0.34	4.09
Grigg–Skjellerup	5.1	1.00	4.94
Tempel 2	5.3	1.36	4.68
Tempel 1	5.5	1.50	4.73
D'Arrest	6.2	1.17	5.61
Pons–Winnecke	6.3	1.25	5.61
Kopff	6.4	1.57	5.34
Schwassmann–Wachmann 2	6.5	2.14	4.83
Giacobini–Zinner	6.5	0.99	5.98
Borrelly	6.8	1.32	5.84
Arend–Rigaux	6.8	1.44	5.76
Brooks 2	6.9	1.84	5.39
Finlay	6.9	1.10	6.19
Daniel	7.1	1.66	5.72
Faye	7.4	1.62	5.98
Wolf	8.4	2.52	5.78
Comas Solá	8.6	1.77	6.59
Wild	13.3	1.98	9.24
Tuttle	13.8	1.02	10.46
Schwassmann–Wachmann 1	15.0	5.45	6.73
Crommelin	27.9	0.74	17.65
Tempel–Tuttle	32.9	0.98	19.56
Olbers	69.5	1.18	32.62
Pons–Brooks	71.0	0.77	33.51
Brorsen-Metcalf	72.0	0.49	34.09
Halley	76.1	0.59	35.33
Swift–Tuttle	120?	?	?

Note: All these have been seen at more than one return apart from Swift–Tuttle, which was observed in 1862 and has so far failed to reappear; it is the parent comet of the Perseid meteors. Few of the periodical comets, apart from Halley's, ever become visible with the naked eye, though some (such as D'Arrest and Brorsen-Metcalf) may occasionally do so.

Appendix 15

Some of the more important Annual Meteor Showers

Shower	Begins	Maximum	Ends	Maximum ZHR	Parent comet
Quadrantids	1 Jan.	3 Jan.	6 Jan.	110	—
Lyrids	19 Apr.	22 Apr.	25 Apr.	20	Thatcher
Eta Aquarids	24 Apr.	6 May	20 May	40	Halley
Capricornids	July	8–26 July	Aug.	6	—
Delta Aquarids	15 July	28 July	20 Aug.	35	—
Perseids	23 July	12 Aug.	18 Aug.	75	Swift–Tuttle
Draconids	10 Oct.	20 Oct.	10 Oct.	variable	Giacobini–Zinner
Orionids	16 Oct.	22 Oct.	26 Oct.	30	Halley
Taurids	20 Oct.	3 Nov.	30 Nov.	12	Encke
Leonids	15 Nov.	17 Nov.	19 Nov.	variable	Tempel–Tuttle
Geminids	7 Dec.	14 Dec.	15 Dec.	70	—
Ursids	17 Dec.	23 Dec.	25 Dec.	12	Tuttle

Note: The radiant of the Quadrantids lies in the constellation of Boötes. All the known parent comets are periodical, though Thatcher's Comet, seen in 1861, has an estimated period of 415 years. The Draconids are usually sparse, but produce occasional rich displays. The Leonids have a general ZHR of no more than 10, but may produce a major display in 1999, as last happened in 1966.

Appendix 16

The Constellations

In the following list, Z = Zodiacal; N = mainly northern hemisphere; S = mainly southern hemisphere; Eq = equatorial; * = too far south to be well seen from Britain or the northern United States; ** = too far north to be well seen from New Zealand or most of South Africa and Australia. Brackets indicate marginal cases.

Constellation	English name	Hemisphere	1st magnitude star(s)
Andromeda	Andromeda	N	
Antlia	The Air-Pump	S*	—
Apus	The Bird of Paradise	S*	—
Aquarius	The Water-Bearer	Z	—
Aquila	The Eagle	N	Altair
Ara	The Altar	S*	—
Aries	The Ram	Z	—
Auriga	The Charioteer	N(**)	Capella
Boötes	The Herdsman	N	Arcturus
Cælum	The Graving Tool	S*	—
Camelopardalis	The Giraffe	N**	—
Cancer	The Crab	Z	—
Canes Venatici	The Hunting Dogs	N**	—
Canis Major	The Great Dog	S	Sirius
Canis Minor	The Little Dog	N	Procyon
Capricornus	The Sea-Goat	Z	—
Carina	The Keel	S*	Canopus
Cassiopeia	Cassiopeia	N**	—
Centaurus	The Centaur	S*	Alpha Centauri, Agena
Cepheus	Cepheus	N**	—
Cetus	The Whale	S	—
Chamæleon	The Chameleon	S*	—
Circinus	The Compasses	S*	—
Columba	The Dove	S(*)	—
Coma Berenices	Berenice's Hair	N**	—
Corona Australis	The Southern Crown	S*	—

Constellation	English name	Hemisphere	1st magnitude star(s)
Corona Borealis	The Northern Crown	N	
Corvus	The Crow	S	—
Crater	The Cup	S	—
Crux Australis	The Southern Cross	S*	Acrux, Beta Crucis
Cygnus	The Swan	N(**)	Deneb
Delphinus	The Dolphin	N	—
Dorado	The Swordfish	S*	—
Draco	The Dragon	N**	—
Equuleus	The Foal	N	—
Eridanus	The River	S(*)	Achernar
Fornax	The Furnace	S	—
Gemini	The Twins	Z	Pollux
Grus	The Crane	S*	—
Hercules	Hercules	N	—
Horologium	The Clock	S*	—
Hydra	The Watersnake	S	—
Hydrus	The Little Snake	S*	—
Indus	The Indian	S*	—
Lacerta	The Lizard	N**	—
Leo	The Lion	Z	Regulus
Leo Minor	The Little Lion	N**	—
Lepus	The Hare	S	—
Libra	The Balance	Z	—
Lupus	The Wolf	S*	—
Lynx	The Lynx	N**	—
Lyra	The Lyre	N(**)	Vega
Mensa	The Table	S*	—
Microscopium	The Microscope	S*	—
Monoceros	The Unicorn	Eq	—
Musca Australis	The Southern Fly	S*	—
Norma	The Rule	S*	—
Octans	The Octant	S*	—
Ophiuchus	The Serpent-bearer	Eq (Z]	—
Orion	Orion	Eq	Rigel, Betelgeux
Pavo	The Peacock	S*	—
Pegasus	The Flying Horse	N	—
Perseus	Perseus	N(**)	—
Phœnix	The Phœnix	S*	—
Pictor	The Painter	S*	—
Pisces	The Fishes	Z	—
Piscis Australis	The Southern Fish	S	Fomalhaut
Puppis	The Poop	S*	—
Pyxis	The Compass	S*	—
Reticulum	The Net	S*	—
Sagitta	The Arrow	N	—

Constellation	English name	Hemisphere	1st magnitude star(s)
Sagittarius	The Archer	Z(*)	—
Scorpius	The Scorpion	Z(*)	Antares
Sculptor	The Sculptor	S	—
Scutum	The Shield	S	—
Serpens	The Serpent	Eq	—
Sextans	The Sextant	S	—
Taurus	The Bull	Z	Aldebaran
Telescopium	The Telescope	S*	—
Triangulum	The Triangle	N	—
Triangulum Australe	The Southern Triangle	S*	—
Tucana	The Toucan	S*	—
Ursa Major	The Great Bear	N**	—
Ursa Minor	The Little Bear	N**	—
Vela	The Sails	S*	—
Virgo	The Virgin	Z	Spica
Volans	The Flying Fish	S*	—
Vulpecula	The Fox	N	—

Appendix 17

Proper Names of Stars

Most of the old proper names of stars have now fallen into disuse, apart from stars of the first magnitude and a few others such as Mizar and Mira. A few stars have more than one name; thus Eta Ursæ Majoris may be 'Benetnasch' and Gamma Virginis 'Porrima'. It is clearly pointless to give all these variations. There are also differences in spelling; Betelgeux may be 'Betelgeuse' or 'Betelgeuze', for example. Alpha Centauri has no official proper name, but has been called 'Toliman' and 'Rigel Kent'; Beta Centauri may be 'Hadar'; the name for Alpha Crucis is often not used, and the name for Beta Crucis. 'Mimosa', is unofficial.

Constellation	Greek letter	Name
Andromeda	Alpha	Alpheratz
	Beta	Mirach
	Gamma	Almaak
	Xi	Adhil
Aquarius	Alpha	Sadalmelik
	Beta	Sadalsuud
	Gamma	Sadachiba
	Delta	Scheat
	Epsilon	Albali
	Theta	Ancha
Aquila	Alpha	Altair
	Beta	Alshain
	Gamma	Tarazed
	Zeta	Dheneb
	Kappa	Situla
	Lambda	Althalimain
Aries	Alpha	Hamal
	Beta	Sheratan

Constellation	Greek letter	Name
	Gamma	Mesartim
	Delta	Boteïn
Auriga	Alpha	Capella
	Beta	Menkarlina
	Zeta	Sadatoni
	Iota	Hassaleh
Boötes	Alpha	Arcturus
	Beta	Nekkar
	Gamma	Seginus
	Epsilon	Izar
	Eta	Saak
	Mu	Alkalurops
	h	Merga
Cancer	Alpha	Acubens
	Gamma	Asellus Borealis
	Delta	Asellus Australis
	Zeta	Tegmine
Canes Venatici	Alpha	Cor Caroli
	Beta	Chara
Canis Major	Alpha	Sirius
	Beta	Mirzam
	Gamma	Muliphen
	Delta	Wezea
	Epsilon	Adhara
	Zeta	Phurad
	Eta	Aludra
Canis Minor	Alpha	Procyon
	Beta	Gomeisa
Caprocornus	Alpha	Al Giedi
	Beta	Dabih
	Gamma	Nashira
	Delta	Deneb al Giedi
	Nu	Alshat
Carina	Alpha	Canopus

Constellation	Greek letter	Name
	Beta	Miaplacidus
	Epsilon	Avior
	Iota	Tureis
Cassiopeia	Alpha	Shedir
	Beta	Chaph
	Delta	Ruchbah
	Epsilon	Segin
	Eta	Achird
	Theta	Marfak
Cepheus	Alpha	Alderamin
	Beta	Alphirk
	Gamma	Alrai
	Xi	Kurdah
Cetus	Alpha	Menkar
	Beta	Diphda
	Gamma	Alkaffaljidhina
	Zeta	Baten Kaitos
	Iota	Deneb Kaitos Shemali
	Omicron	Mira
Columba	Alpha	Phakt
	Beta	Wazn
Coma Berenices	Alpha	Diadem
Corona Borealis	Alpha	Alphekka
	Beta	Nusakan
Corvus	Alpha	Alkhiba
	Beta	Kraz
	Gamma	Minkar
	Delta	Algorel
Crater	Alpha	Alkes
Crux	Alpha	Acrux
Cygnus	Alpha	Deneb
	Beta	Albireo

Constellation	Greek letter	Name
	Gamma	Sadr
	Epsilon	Gienah
	Pi	Azelfafage
	Omega	Ruchba
Delphinus	Alpha	Svalocin
	Beta	Rotanev
Draco	Alpha	Thuban
	Beta	Alkwaid
	Gamma	Eltamin
	Delta	Taïs
	Epsilon	Tyl
	Zeta	Aldhibah
	Eta	Aldhibain
	Iota	Edasich
	Lambda	Giansar
	Mu	Alrakis
	Xi	Juza
	Psi	Dziban
Equuleus	Alpha	Kitalpha
Eridanus	Alpha	Achernar
	Beta	Kursa
	Gamma	Zaurak
	Delta	Rana
	Zeta	Zibal
	Eta	Azha
	Theta	Acamar
	Omicron[1]	Beid
	Omicron[2]	Keid
	Tau	Angetenaar
	53	Sceptrum
Gemini	Alpha	Castor
	Beta	Pollux
	Gamma	Alhena
	Delta	Wasat
	Epsilon	Mebsuta
	Zeta	Mekbuda
	Eta	Propus

Constellation	Greek letter	Name
	Mu	Tejat
	Xi	Alzirr
Grus	Alpha	Alnair
	Beta	Al Dhanab
Hercules	Alpha	Rasalgethi
	Beta	Kornephoros
	Delta	Sarin
	Zeta	Rutilicus
	Kappa	Marsik
	Lambda	Masym
	Omega	Cujam
Hydra	Alpha	Alphard
Leo	Alpha	Regulus
	Beta	Denebola
	Gamma	Algieba
	Delta	Zosma
	Epsilon	Asad Australis
	Zeta	Adhafera
	Theta	Chort
	Lambda	Alterf
	Mu	Rassalas
	Omicron	Subra
Leo Minor	46	Præcipua
Lepus	Alpha	Arneb
	Beta	Nihal
Libra	Alpha	Zubenelgenubi
	Beta	Zubenelchemale
	Gamma	Zubenelhakrabi
	Sigma	Zubenalgubi (= Gamma Scorpii)
Lynx	31	Alsciaukat
Lyra	Alpha	Vega
	Beta	Sheliak
	Gamma	Sulaphat

Constellation	Greek letter	Name
	Eta	Aladfar
	Mu	Al Athfar
Ophiuchus	Alpha	Rasalhague
	Beta	Cheleb
	Delta	Yed Prior
	Epsilon	Yed Post
	Zeta	Han
	Eta	Sabik
Orion	Alpha	Betelgeuse
	Beta	Rigel
	Gamma	Bellatrix
	Delta	Mintaka
	Epsilon	Alnilam
	Zeta	Alnitak
	Eta	Algjebbah
	Kappa	Saiph
	Lambda	Heka
	Upsilon	Thabit
Pegasus	Alpha	Markab
	Beta	Scheat
	Gamma	Algenib
	Epsilon	Enif
	Zeta	Homan
	Eta	Matar
	Theta	Biham
	Mu	Sadalbari
Perseus	Alpha	Mirphak
	Beta	Algol
	Zeta	Atik
	Kappa	Misam
	Xi	Menkib
	Omicron	Ati
	Tau	Kerb
	Upsilon	Nembus
Phœnix	Alpha	Ankaa
Pisces	Alpha	Al Rischa
	Eta	Alpherg

Constellation	Greek letter	Name
Piscis Australis	Alpha	Fomalhaut
	Beta	Fum al Samakah
Puppis	Zeta	Suhail Hadar
	Xi	Asmidiske
	Rho	Turais
Sagittarius	Alpha	Rukbat
	Beta	Arkab
	Gamma	Alnasr
	Delta	Kaus Meridionalis
	Epsilon	Kaus Australis
	Zeta	Ascella
	Lambda	Kaus Borealis
	Pi	Albaldah
	Sigma	Nunki
Scorpius	Alpha	Antares
	Beta	Graffias
	Delta	Dschubba
	Epsilon	Wei
	Theta	Sargas
	Kappa	Girtab
	Lambda	Shaula
	Nu	Jabbah
	Sigma	Alniyat
	Upsilon	Lesath
	Omega	Jabhat al Akrab
Serpens	Alpha	Unukalhai
	Theta	Alya
Taurus	Alpha	Aldebaran
	Beta	Alnath
	Gamma	Hyadum Primus
	Epsilon	Ain
	Eta	Alcyone
	17	Electra
	19	Taygete
	20	Maia
	21	Asterope
	23	Merope
	27	Atlas
	28	Pleione

Constellation	Greek letter	Name
Triangulum	Alpha	Rasalmothallah
Triangulum Australe	Alpha	Atria
Ursa Major	Alpha	Dubhe
	Beta	Merak
	Gamma	Phad
	Delta	Megrez
	Epsilon	Alioth
	Zeta	Mizar
	Eta	Alkaid
	Iota	Talita
	Kappa	Al Kaprah
	Lambda	Tania Borealis
	Mu	Tania Australis
	Nu	Alula Borealis
	Xi	Alula Australis
	Omicron	Muscida
	Pi	Ta Tsun
	Chi	Alkafazah
	80	Alcor
Ursa Minor	Alpha	Polaris
	Beta	Kocab
	Gamma	Pherkad Major
	Delta	Yildun
	Zeta	Alifa
	Eta	Alasco
Vela	Gamma	Regor
	Kappa	Markeb
	Lambda	Al Suhail al Wazn
Virgo	Alpha	Spica
	Beta	Zavijava
	Gamma	Arich
	Delta	Minelauva
	Epsilon	Vindemiatrix
	Zeta	Heze
	Eta	Zaniah
	Iota	Syrma
	Lambda	Khambalia

Appendix 18

Stars of the First Magnitude

Conventionally, stars of the 'first magnitude' are taken as being those above magnitude 1.5. The values given here follow the Cambridge *Sky Catalogue 2000*.

Star	Proper name	Magnitude	Luminosity, (Sun = 1)	Spectrum	Distance: (light years)
Alpha Canis Majoris	Sirius	−1.46	26	A1	8.6
Alpha Carinæ	Canopus	−0.72	200000	F0	1200
Alpha Centauri	—	−0.27	1.5	K1 + G2	4.3
Alpha Boötis	Arcturus	−0.04	115	K2	36
Alpha Lyræ	Vega	0.03	52	A0	26
Alpha Aurigæ	Capella	0.08	70	G8	42
Beta Orionis	Rigel	0.12	60000	B8	900
Alpha Canis Minoris	Procyon	0.38	11	F5	11.4
Alpha Eridani	Achernar	0.46	780	B5	85
Alpha Orionis	Betelgeuse	variable	15000	M2	310
Beta Centauri	Agena	0.61	10500	B1	460
Alpha Aquilæ	Altair	0.77	10	A7	16.6
Alpha Crucis	Acrux	0.83	3200 + 2000	B1 + B3	360
Alpha Tauri	Aldebaran	0.85	100	K5	68
Alpha Scorpii	Antares	0.96	7500	M1	330
Alpha Virginis	Spica	0.98	2100	B1	260
Beta Geminorum	Pollux	1.14	60	K0	36
Alpha Piscis Australis	Fomalhaut	1.16	13	A3	22
Alpha Cygni	Deneb	1.25	70000	A2	1800
Beta Crucis	—	1.25	8200	B0	425
Alpha Leonis	Regulus	1.35	130	B7	85

Appendix 19

Standard Stars for Each Magnitude

It may be helpful to know the magnitudes of a few standard stars for each magnitude. In choosing this list, I have tried to satisfy observers in each hemisphere!

1	Alpha Tauri (Aldebaran)	0.85
	Alpha Scorpii (Antares)	0.98
	Alpha Virginis (Spica)	0.98
	Beta Geminorum (Pollux)	1.14
$1\frac{1}{2}$	Epsilon Canis Majoris	1.50
	Alpha Geminorum (Castor)	1.58
	Lambda Scorpii	1.63
	Gamma Crucis	1.63
2	Beta Canis Majoris	1.98
	Alpha Ursæ Minoris (Polaris)	1.99
	Alpha Arietis	2.00
	Sigma Sagittarii	2.02
	Beta Ceti	2.04
$2\frac{1}{2}$	Gamma Ursæ Majoris (Phad)	2.44
	Epsilon Cygni	2.46
	Alpha Pegasi	2.49
	Zeta Centauri	2.55
3	Zeta Aquilæ	2.99
	Zeta Tauri	3.00
	Epsilon Corvi	3.00
	Gamma Gruis	3.01
	Gamma Ursæ Minoris	3.05
$3\frac{1}{2}$	Tau Ceti	3.50
	Beta Boötis	3.50
	Beta Cancri	3.52
	Iota Cephei	3.52
	Delta Eridani	3.54

4	Omega Piscium	4.01
	Theta Draconis	4.01
	Nu Virginis	4.03
	Alpha Chamæleontis	4.07
$4\frac{1}{2}$	Delta Equulei	4.49
	Lambda Virginis	4.52
	Nu Andromedæ	4.53
	Beta Piscium	4.53
	Beta Crateris	4.45
5	Eta Ursæ Minoris	4.95
	Kappa Aquilæ	4.95
	Omega Cassiopeiæ	4.99
	Zeta Sagittæ	5.00
	Lambda Telescopii	5.00
$5\frac{1}{2}$	Theta Ursæ Minoris	5.33
	Rho Coronæa Borealis	5.43
	Epsilon Trianguli	5.44
	3 Piscis Australis	5.55

Appendix 20

The Greek Alphabet

α	Alpha	ν	Nu
β	Beta	ξ	Xi
γ	Gamma	o	Omicron
δ	Delta	π	Pi
ϵ	Epsilon	ρ	Rho
ζ	Zeta	σ	Sigma
η	Eta	τ	Tau
θ	Theta	υ	Upsilon
ι	Iota	ϕ	Phi
κ	Kappa	χ	Chi
λ	Lamba	ψ	Psi
μ	Mu	ω	Omega

Appendix 21

Stellar Spectra

Type	Surface Temperature (°C)[b]	Colour	Typical star(s)	Remarks
W	Up to 80000	White	Gamma Velorum	Wolf–Rayet stars. Many bright lines. Rare.
O	35000–40000	White	Zeta Orionis	Bright and dark lines; helium prominent. Rare.
B	12000–25000	White or bluish	Spica, Rigel	Helium prominent
A	7500–11000	White	Sirius, Vega	Hydrogen prominent.
F	6000–7500	Yellowish	Canopus, Polaris	Calcium prominent. Visually often look white.
G	5000–6000	Yellow	Capella (giant), Sun (dwarf)	Metallic lines numerous.
K	3500–5000	Orange	Arcturus (giant), Tau Ceti (dwarf)	Metallic lines strong, hydrogen weak.
M	3000–3500	Orange-red	Betelgeux (giant), Proxima (dwarf)	Broad titanium oxide and calcium bands.
R	2300[a]	Reddish	S Apodis	Carbon prominent.
N	2500[a]	Red	R Leporis	Carbon prominent.
S	2600	Red	Chi Cygni, R Andromedæ	Zirconium oxide and titanium oxide prominent.

Note:
[a] Today, R and N stars are often classed together as Type C.
[b] Temperatures here are given in degrees Centigrade (°C). To convert to degrees Fahrenheit (°F) (or vice versa):

$$°F = (°C \times 1.8) + 32$$
$$°C = (\tfrac{5}{9}) \times (°F - 32)$$

For large values, of course, the '32' correction can be ignored. For absolute temperatures (K), or degrees Kelvin. To convert from K to C, subtract 273. To convert from C to K, add 273.

Appendix 22

Limiting Magnitudes and Separations for Various Apertures

It is almost impossible to give definite values for limiting magnitudes and separations, since so much must depend upon individual observers. The following table is approximate only, and based largely on my own experience: others will no doubt disagree!

The third column refers to stars of equal brilliancy and of about the sixth magnitude. (Where the components are unequal, the double will naturally be a more difficult object, particularly if one star is much brighter than the other.) The theoretical limit is generally defined by Dawes' Limit: $r = 4.56/d$, where r is the separation in seconds of arc and d is the diameter of the telescope object-glass. Thus for a 12-inch telescope, the Dawes' Limit would be 0.38". In this table I have done my best to be realistic.

Aperture (inches)	Faintest magnitude	Smallest separation (seconds of arc)
2	10.5	2.5
3	11.4	1.8
4	12.0	1.5
5	12.5	1.0
6	12.9	0.9
7	13.2	0.7
8	13.5	0.6
10	14.0	0.5
12	14.4	0.4
15	14.9	0.3

Appendix 23

Angular Distances between selected Stars: and Declinations of Bright Stars

It may be useful to give the angular distances between some selected stars, as this will help observers unused to angular measurement. The distance round the horizon is of course 360 degrees, and from the horizon to the zenith 90 degrees; the Sun and Full Moon have angular diameters of about half a degree, which is the same as that of a one-inch disk held at $9\frac{1}{2}$ feet from the eye.

Approximate separation	Stars
60	Polaris to Pollux: Alpha Ursæ Majoris to Beta Cassiopeiæ
50	Sirius to Castor: Polaris to Vega
45	Polaris to Deneb: Spica to Antares
40	Capella to Betelgeuse: Castor to Regulus
35	Vega to Altair: Capella to Pollux
30	Polaris to Beta Cassiopeiæ: Aldebaran to Capella
25	Sirius to Procyon: Vega to Deneb
20	Betelgeuse to Rigel: Procyon to Pollux
15	Alpha Centauri to Acrux: Alpha Andromedæ to Beta Andromedæ
10	Betelgeuse to Delta Orionis: Acrux to Agena
5	Alpha Ursæ Majoris to Beta Ursæ Majoris
$4\frac{1}{2}$	Castor to Pollux: Alpha Centauri to Agena
3	Delta Scorpii to Beta Scorpii
$1\frac{1}{2}$	Altair to Beta Aquilæ
$1\frac{1}{2}$	Beta Arietis to Gamma Arietis
1	Atlas to Electra (Pleiades)

To find the diameter of a telescopic field, select some star very near the celestial equator (such as Delta Orionis or Zeta Virginis) and allow it to drift right through the field. This time in minutes and seconds, multiplied by 15, will give the angular diameter of the field in minutes and seconds of arc. For instance, if Delta Orionis takes

1 minute 3 seconds to pass through the field, the diameter is 1 minute 3 seconds × 15 = 15′45″ of arc.

The following table gives the declinations of some bright stars. You can then work out whether they are or not visible from your own latitude, as described on page 135.

Star	Declination (degrees)	Constellation
Alpheratz	+ 29	Andromeda
Altair	+ 90	Aquila
Capella	+ 46	Auriga
Arcturus	+ 19	Boötes
Sirius	− 17	Canis Major
Procyon	+ 5	Canis Minor
Canopus	− 53	Carina
Shedir	+ 57	Cassiopeia
Alpha Centauri	− 61	Centaurus
Mira	− 03	Cetus
Acrux	− 63	Crux
Gamma Crucis	− 57	Crux
Deneb	+ 45	Cygnus
Achernar	− 57	Eridanus
Castor	+ 32	Gemini
Alnair	− 47	Grus
Regulus	+ 12	Leo
Vega	+ 39	Lyra
Rigel	− 08	Orion
Alpha Pavonis	− 57	Pavo
Markab	+ 15	Pegasus
Algol	+ 41	Perseus
Fomalhaut	− 30	Piscis Australis
Zeta Puppis	− 40	Puppis
Kaus Australis	− 34	Sagittarius
Antares	− 26	Scorpius
Shaula	− 37	Scorpius
Aldebaran	+ 17	Taurus
Alpha Trianguli Australe	− 69	Triangulum Australe
Dubhe	+ 62	Ursa Major
Alkaid	+ 49	Ursa Major
Kocab	+ 74	Ursa Minor
Regor	− 47	Vela
Spica	− 11	Virgo

Appendix 24

Double Stars for Various Apertures

Again, everything depends upon conditions and the individual skill of the observer; but the following double stars ought to be separable with telescopes of the aperture indicated. If you cannot separate, say, Beta Cygni and Mizar with a 1in, there is something wrong.

Aperture	Star	Position angle (degrees)	Separation (seconds)	Magnitudes
1	Zeta Ursæ Majoris (Mizar)	152	14.4	2.3, 4.0
	Beta Cygni (Albireo)	054	34.4	3.1, 5.1
	Alpha Crucis (Acrux)	115	4.4.	1.4, 1.9
	Alpha Herculis	107	4.7	3 variable 5.4
	Gamma Velorum	220	41.2	1.9, 4.2
	61 Cygni	148	30.0	5.2, 6.0
	Alpha Centauri	215	19.7	0.0, 1.2
2	Gamma Leonis	124	4.3	2.2, 3.5
	Epsilon Boötis	339	2.8	2.5, 4.9
	Eta Cassiopeiæ	293	12.0	3.4, 7.5
	Alpha Ursæ Minoris (Polaris)	218	18.4	2.0, 9.0
3	Theta Virginis	343	7.1	4.4, 9.4
	Zeta Coronæ Borealis	305	6.3	5.1, 6.0
4	Theta Aurigæ	313	3.6	2.6, 7.1
	Iota Ursæ Majoris	016	406.7	3.1, 10.2
	Delta Cygni	137	2.4	2.9, 6.3
	Eta Orionis	080	1.5	3.8, 4.8
5	Zeta Boötis	303	1.0	4.5, 4.6
	Eta Coronæ Borealis	024	1.0	5.6, 5.9
8	Lambda Cassiopeiæ	176	0.6	5.5, 5.8
	Gamma Andromedæ	103	0.5	5.5, 6.3
	Omega Leonis	053	0.5	5.9, 6.5

Appendix 25

The Observation of Variable Stars

Variable star work is becoming more and more popular among amateurs. Some notes on it have already been given, but it may be as well to summarize them here, even though it means a certain amount of duplication. To present a full account would need a complete book to itself, particularly as there are so many variables within the range of a small telescope, so that I have confined myself to a few 'typical cases' – largely, I admit, to stars which are on my own observational list.

The main classes are:

ECLIPSING VARIABLES

1. Algol stars. One component much brighter than the other, producing one deep minimum and a secondary minimum which is generally too small to be detected without sensitive equipment.
2. Beta Lyræ stars. Components less unequal, and very close together; minima unequal, but both very marked.
3. W Ursæ Majoris stars. Close binaries; short periods, less than 12 hours. No really bright example.

To study eclipsing binaries properly needs photoelectric equipment. True, visual observation is not to be despised, and there are some eclipsing stars which will repay attention; but I have never tackled them myself, which is an excellent reason for saying no more about them here.

PULSATING VARIABLES

Short period

1. RR Lyræ stars. Very regular; very short periods; common in globular clusters, though many of them (including RR Lyræ itself) are not cluster members. No bright examples.
2. Cepheids, such as Delta Cephei, Eta Aquilæ and Beta Doradûs. Some have a

considerable range in magnitude; others change very little. The changes are regular; though some useful work can be done visually, photoelectric equipment is desirable. Classical Cepheids belong to Population I.

3. Type II Cepheids, or W Virginis stars. These have a different period—luminosity law. The most celebrated example is the southern Kappa Pavonis.

There are various other classes of short-period stars, but I do not propose to discuss them here, because they are not suited to visual observation.

Longer period

1. Mira stars, named after Mira Ceti. Both period and range alter, and the light-curve is never repeated exactly from one cycle to the next. Most of them are red giants. They are ideal for amateur observation; it is usually quite good enough to estimate their magnitudes down to 0.1.
2. Semi-regular stars. Smaller ranges, and periods which are usually shorter than with the Mira variables; but the periods are very rough indeed, and are subject to interruption, so that amateur observation is very valuable.
3. RV Tauri stars. Alternate deep and shallow minima, but the light-curves are never repeated exactly, and the behaviour is often quite irregular for a while. R Scuti is the brightest example.

Irregular

This is a general term; some stars which are classed as irregular may in fact be semi-regular, and it is often difficult to decide. There are also some stars which simply do not seem to fit in anywhere; Rho Cassiopeiæ is one.

ERUPTIVE VARIABLES

1. SS Cygni or U Geminorum stars. Nova-like outbursts at mean intervals which range from 20 to 600 days, but are never predictable. Ideal for amateur observation, but most of them are so faint that they require large apertures.
2. R Coronæ Borealis stars. These remain at or near maximum for most of the time, but show sudden, unpredictable drops to minimum. Observation of them is very useful — when a fall begins, an amateur may well be the first to note it. Only R Coronæ itself is ever visible with the naked eye.
3. T Tauri stars. Rapid, irregular fluctuations; these are faint, young stars.
4. Z Camelopardalis stars. Similar to SS Cygni stars except that at unpredicatble intervals the variations cease for a while, and there is a 'standstill'. Rare; large apertures needed.
5. Flare stars, such as UV Ceti and AD Leonis. These show sudden rises, which may amount to several magnitudes; the outburst takes only a few minutes, and the subsequent fading may take hours. Again, most of them are faint, and the technique is different from that used for other variables; the star is kept under

observation for a set period. The easiest example is AD Leonis, which is in the same field with Gamma Leonis; its normal magnitude is 9.5, but it can flare up to above 9. UV Ceti, the prototype, is usually 12.9, but on one occasion rose abruptly to 5.9. All these stars are nearby red dwarfs. Observation of them is fascinating, but is a matter for the real specialist with endless patience; if the work is to be carried out visually rather than with recording equipment, it is best to work with a team, each member taking relatively short spells at the eyepiece.

6. P Cygni stars. Slow, erratic variations; possibly related to novæ. All are extremely hot, luminous, unstable and remote.

7. Novæ. Rapid rises, followed by a slower decline.

When estimating the magnitude of a variable, always use several comparison stars – if you can. Either the step-method or the fractional may be employed (see pp. 167–8). (I use the step, though many people tell me that the fractional is slightly more accurate.) What usually happens, of course, is that a discrepancy is found. Suppose you estimate the variable as 0.3 below comparison star A, and 0.5 above B; on looking up your charts you find that A is 7.0 and B is 7.6, in which case the variable would work out at 7.3 if you rely upon A and 7.1 if you prefer B. By using three or more comparison stars a good figure can usually be obtained, but odd things can happen sometimes. Moreover, it is not easy to compare a red star with a white one, and many long-period variables are red. U Cygni, for example, is intensely red and is notoriously difficult to estimate correctly, as I know to my cost.

Naked-eye estimates are quite valid for bright variables, but take care to allow for 'extinction', when the variable and the comparison star are at different altitudes. The closer a star is to the horizon, the more of its light will be lost. Table A25.1 gives the amount of extinction for various altitudes above the horizon. Above an altitude of 45 degrees, extinction can be neglected for most practical purposes. For a telescopic or binocular variable extinction need not be taken into account, as the variable and the comparison stars will be at virtually the same altitude.

Table A25.1. *The amount of extinction for various altitudes above the horizon*

Altitude (degrees)	Extinction in magnitudes
1	3
2	2.5
4	2
10	1
13	0.8
15	0.7
17	0.6
21	0.4
26	0.3
32	0.2
43	0.1

The following notes and charts are specimens only. Anyone who is interested can obtain others, provided that he is a member of a society such as the British Astronomical Association, the Royal Astronomical Society of New Zealand or the American Association of Variable Star Observers. Obviously, one can tackle as many (or as few) stars as one likes; I have 35 on my own list, but really enthusiastic observers concentrate upon dozens, while I admit to being mainly a lunar and planetary observer.

BINOCULAR VARIABLES

Many interesting variables are within the range of binoculars. If only one pair is available, I recommend 7 × 50, but obviously a larger aperture will give a greater light-grasp.

R Lyræ. Semi-regular; 3.9–5.0. The period is said to be about 46 days. An awkward star, because there are no suitable comparisons close by; put R Lyræ on one side of the binocular field, and Eta will be almost on the far side. Comparisons: Eta (4.4), Theta (4.4), 16 (5.1) (Fig. A25.1). Actually, I always make Theta a little brighter than Eta, and since it is a K-type star it may well be slightly variable.

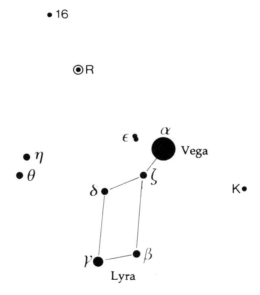

Fig. A25.1. Comparison stars for R Lyræ.

U and EU Delphini. Easy to find. U has been listed as a semi-regular, but I have never found any real trace of a period; range 7.6–8.9. EU (5.8–6.9) also seems irregular to me. Comparisons: A (5.6), B (6.3), D (6.3), H (6.8) and G (7.3) (Fig. A25.2).

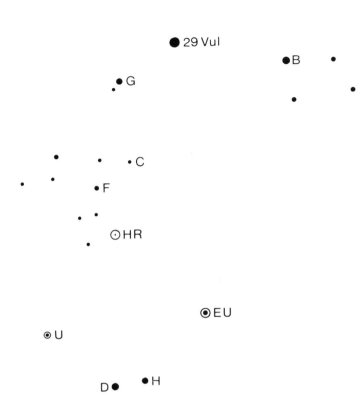

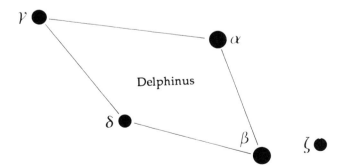

Fig. A25.2. *Comparison stars for U and EU Delphini.*

W Cygni. A very interesting star. The range is 5.0–7.6, and it is officially classed as a semi-regular with a period of 126 days, but I have never been able to confirm this. To locate it, find Rho Cygni (Fig. A25.3). The field will be instantly recognizable in binoculars. Comparisons: 75 Cygni (5.0), D (5.4), A (6.1), B (6.7), K (6.8), L (7.5) (Fig. A25.3). Beware of the star marked X, which is itself variable by about half a magnitude. (Note that the letters in these charts are those used by variable star observers, and are not 'official'; for instance, X Cygni, near Lambda, is a Cepheid, and is not the same as the X on this chart.) Note 75 Cygni; I will return to it later, as it is the guide-star to the famous eruptive variable SS Cygni.

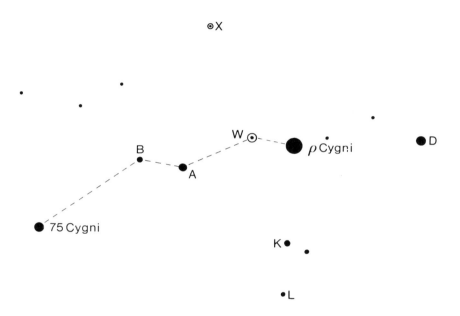

Fig. A25.3. Comparison stars for W Cygni.

R Scuti. The brightest of the RV Tauri stars. Range 4.4–8.2; rough period 140 days. It is easily found from Lambda Aquilæ and the famous 'Wild Duck' cluster, M.11 (Fig. A25.4). Comparisons: A (4.5), B (4.8), C (5.0), D (5.2), E (5.6), F (6.1), G (6.8), H (7.1), K (7.7) (Fig. A25.4). R makes a well-marked quadrilateral with F, G and H.

X Persei. Range 6–7; no period is known. It is easily found, near Zeta Persei. Comparisons: 40 (5.0), 42 (5.1), Y (5.7), A (6.1), B (6.2), C (6.6) (Fig. A25.5). X Persei is a source of X-radiation.

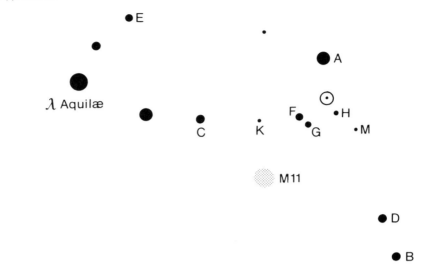

Fig. A25.4. *Comparison stars for R Scuti.*

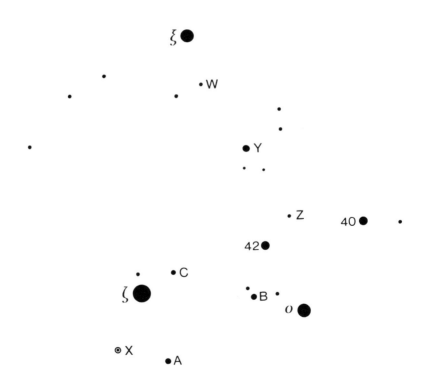

Fig. A25.5. *Comparison stars for X Persei.*

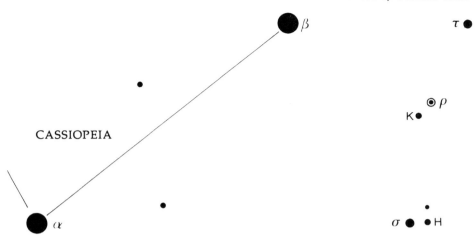

Fig. A25.6. Comparison stars for Rho Cassiopeiæ.

Rho Cassiopeiæ. Close to Beta (Fig. A25.6). An ideal binocular object. Its usual magnitude is about 5, but it has been know to fall below 6; these minima are infrequent – I have yet to see one – and nobody knows what sort of variable it is. Comparisons: Sigma (4.9), Tau (5.1), H (5.7), K (6.1) (Fig. A25.6). Rho Cassipeiæ is a very remote, luminous supergiant.

These are only a sample of the many variables which may be followed with binoculars. Also, some long-period stars, such as U Orionis, R Leonis and R Serpentis, are binocular objects when near maximum; and some, notably Mira itself and Chi Cygni, reach naked-eye visibility.

TELESCOPIC VARIABLES

There are so many variables within the range of, say, a 6-inch telescope that no observer can hope to deal with them all. The following specimen charts are inverted, for telescopic use by northern-hemisphere observers.

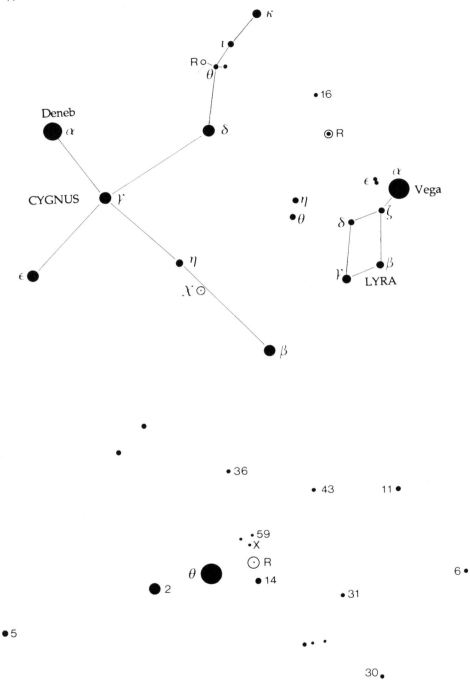

Fig. A25.7(a) Finder charts for R Cygni and Chi Cygni (R Lyræ is also shown); (b) Telescopic field for R Cygni, with comparison stars.

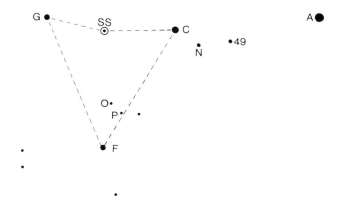

Fig. A25.8. Telescopic field of SS Cygni, with comparison stars.

R Cygni. This is extremely easy to find, since it lies in the field with the 4th-magnitude star Theta Cygni – just off the map on page 304, but shown in Fig. A25.7 *a* & *b*. It has range of from 6.1 to 14.2, and a period of 426 days. Comparison star 2 (6.6) is in the field with Theta. Other comparisons are 5 (9.0), 14 (9.9), 31 (11.0), 36 (11.4), 43 (11.9), 59 (12.3), and x (12.8). This is not a full sequence, but is is enough to show the way in which R Cygni behaves. It is of type S, and is very red. Of course, it passes below the range of small telescopes when faint.

SS Cygni. This is an excellent example of a fainter variable which is easy to find (Fig. A25.8). Locate 75 Cygni, near Rho, which is identifiable because it is distinctly red. Then look for the triangle made up of C (8.5), G (9.6) and F (9.4); SS lies between C and G. Also to hand are A (8.0), N and 49 (each 11.3), O (11.8), P (12.1). At its usual brightness SS is comparable with O; at its best it can equal A. The average period between outbursts is 50 days, but this *is* only an average.

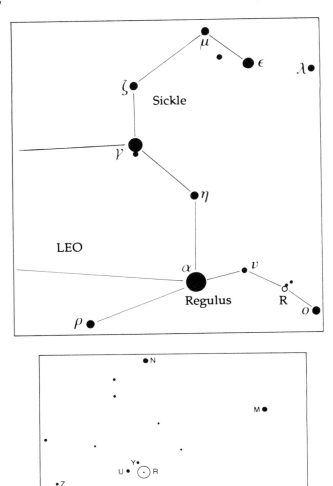

Fig. A25.9(a) Finder chart for R Leonis; (b) Telescopic field for R Leonis, with comparison stars.

R. Leonis. Range 4.4–11.3, though at most maxima the magnitude remains below 5. Mira type. It lies near Regulus, and makes up a trio with 18 Leonis (5.8) and 19 (6.4) (Fig. A25.9*a*). Other comparisons are 21 Leonis (6.6), M (7.2), N (7.5), Q (8.2), U (9.0), Y (9.6) and Z (10.1) (Fig. A25.9*b*). The inset chart – not inverted this time – shows the trio together with Regulus and the fourth-magnitude Omega Leonis. R Leonis is a convenient star for practice, because it never becomes very faint.

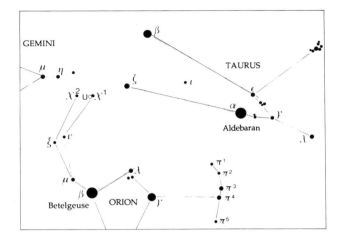

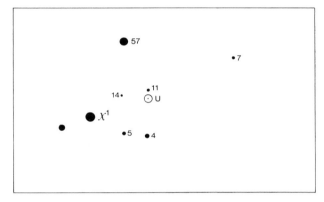

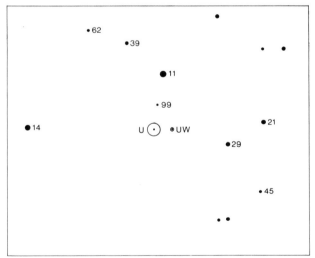

Fig. A25.10(a) *Finder field for U Orionis; (b) telescopic field of U Orionis; and (c) more detailed field for U Orionis, showing fainter comparison stars.*

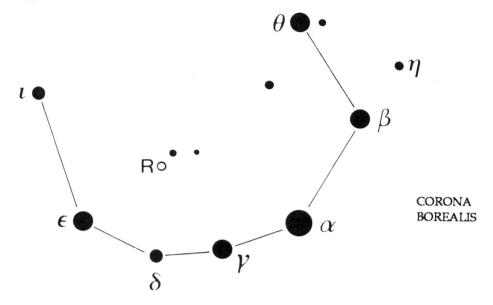

Fig. A25.11. Finder field for R Coronæ Borealis.

U Orionis. Mira type; 4.8–12.6; 372 days. Most maxima remain below magnitude 5. Start from Zeta Tauri (page 290) and locate the pair of stars Chi[1] and Chi[2] Orionis; from these, identify the star 11 (8.9) (Fig A25.10*a* & *b*). Figure A25.10*c* (inverted for telescopic use) shows the field round 11. Comparisons: Chi[1] (4.5), Chi[2] (5.8), 4 (7.2), 5 (7.9), 7 (8.4), 11 (8.9), 14 (9.2), 21 (9.7), 29 (9.9), 39 (10.6), 45 (11.2), 62 (11.6), 99 (12.3). Beware of UW Orionis, which is a Beta Lyræ eclipsing binary with a range of from 10.9 to 11.8 and a period of 1 day.

R Coronæ. Range 5.8–15. When at maximum – that is to say, for most of the time – it is on the fringe of naked-eye visibility, and is a binocular object; compare it with M (6.6) or 1 (7.2) (Fig. A25.11). If you look for it with binoculars and cannot find it, you may be sure that it has suffered one of its deep, unpredictable falls, in which case you will need a large telescope and a set of specialist charts to locate it.

In Table A25.2 I have selected some variable stars which are suited for amateur observation. The list makes no pretence of being in any way complete – far from it – but if you choose some of these stars and follow them, with charts obtainable from an astronomical society, you will soon find that variables are fascinating companions.

Not all these fainter variables are shown on the maps given here (Figs A25.1–25.11). I have therefore added the right ascensions and declinations, so that you can locate them on large-scale charts.

Table A25.2. *Variable stars suited for amateur observations*

Star	Right ascension hours	Right ascension minutes	Delination degrees	Delination minutes	Range	Type	Period (d)	Spectrum
R Andromedæ	00	24.0	+38	35	5.8–14.9	Mira	409	M
U Antliæ	10	35.2	−39	34	5.7–6.8	Irregular	—	N
Theta Apodis	14	05.3	−76	48	6.4–8.6	Semi-regular	119	M
S Apodis	15	04.3	−71	53	9.5–15	R Coronæ	—	R
R Aquarii	23	43.8	−15	17	5.8–12.4	Z Andr	387	M + Pec.,
Eta Aquilæ	19	52.5	+01	00	3.5–4.4	Cepheid	7.2	F–G
R Aquilæ	19	06.4	+08	14	5.5–12.0	Mira	284	M
R Aræ	16	39.7	−57	00	6.0–6.9	Algol	4.48	B
U Arietis	03	11.0	+14	48	7.2–15.2	Mira	322	M
R Arietis	02	16.1	+25	03	7.4–13.7	Mira	187	M
R Boötis	14	37.2	+26	44	6.2–13.1	Mira	223	M
R Camelopardalis	14	17.8	+83	50	7.0–14.4	Mira	270	S
R Cancri	08	16.6	+11	44	6.1–11.8	Mira	362	M
Y Canum Venatic.	12	45.1	+45	26	7.4–10.0	Semi-regular	157	N
R Canis Majoris	07	19.5	−16	24	5.7–6.3	Algol	1.2	F
S Canis Minoris	07	32.7	+08	19	6.6–13.2	Mira	333	M
R Carinæ	09	32.2	−62	47	3.9–10.5	Mira	309	M
S Carinæ	10	09.4	−61	33	4.5–9.9	Mira	149	K–M
U Carinæ	10	57.8	−59	44	5.5–7.0	Cepheid	38.8	F–G
ZZ Carinæ (l Car)	09	45.2	−62	30	3.3–4.2	Cepheid	35.5	F–G
Gamma Cassiopeiæ	00	56.7	+60	43	1.6–3.3	Irregular	—	B
Rho Cassiopeiæ	23	54.4	+58	30	4.1–6.2	?	—	F
R Cassiopeiæ	23	58.4	+51	24	4.7–13.5	Mira	431	M
S Cassiopeiæ	01	19.7	+72	37	7.9–16.1	Mira	612	S
T Cassiopeiæ	00	23.2	+55	48	6.9–13.0	Mira	445	M
Mu Centauri	13	49.6	−42	28	2.9–3.5	Irregular	—	B
R Centauri	14	16.6	−59	55	5.3–11.8	Mira	546	M
S Centauri	12	24.6	−49	26	6.0–7.0	Semi-regular	65	N
T Centauri	13	41.8	−33	36	5.5–9.0	Semi-regular	60	K–M
X Centauri	11	49.2	−41	45	7.0–13.8	Mira	315	M
Delta Cephei	22	29.2	+58	25	3.5–4.4	Cepheid	5.4	F–G
Mu Cephei	21	43.5	+58	47	3.4–5.1	Irregular	—	M
S Cephei	21	35.2	+78	37	7.4–12.9	Mira	487	N
T Cephei	21	09.5	+68	29	5.2–11.3	Mira	388	M
Omicron Ceti	02	19.3	−02	59	1.7–10.1	Mira	332	M
T Ceti	00	21.8	−20	03	5.0–6.9	Semi-regular	159	M
U Ceti	02	33.7	−13	09	6.8–13.4	Mira	235	M
UV Ceti	01	38.8	−17	58	6.8–13.0	Flare	—	M
R Chamæleontis	08	21.8	−76	21	7.5–14.2	Mira	335	M
R Columbæ	05	50.5	−29	12	7.8–15.0	Mira	328	M
R Comæ Berenices	12	04.0	+18	49	7.4–14.6	Mira	363	M
R Coronæ Bor.	15	48.6	+28	09	5.7–15	R Coronæ	—	F
S Coronæ Bor.	15	21.4	+31	22	5.8–14.1	Mira	360	M
T Coronæ Bor.	15	59.5	+25	55	2.0–10.8	Rec. nova	—	M + Q
U Coronæ Bor.	15	18.2	+31	39	7.7–8.8	Algol	13.5	B + F
R Corvi	12	19.6	−19	15	6.7–14.4	Mira	317	M
R Crucis	12	23.6	−61	38	6.4–7.2	Cepheid	5.8	F–G
S Crucis	12	54.4	−58	26	6.2–6.9	Cepheid	4.69	F–G
T Crucis	12	21.4	−62	17	6.3–6.8	Cepheid	6.73	F
Chi Cygni	19	50.6	+32	55	3.3–14.2	Mira	407	S
P Cygni	20	17.8	+38	02	3–6	Irregular	—	B pec.
R Cygni	19	36.8	+50	12	6.1–14.2	Mira	426	M

Table A25.2. (cont.)

Star	Right ascension hours	Right ascension minutes	Delination degrees	Delination minutes	Range	Type	Period (d)	Spectrum
U Cygni	20	19.6	+ 47	54	5.9–12.1	Mira	462	N
W Cygni	21	36.0	+ 45	22	5.0–7.6	Semi-regular	126	M
X Cygni	20	43.4.	+ 35	35	5.9–6.9	Cepheid	16.4	F–G
SS Cygni	21	42.7	+ 43	35	8.4–12.4	SS Cygni	± 50	A–G
U Delphini	20	45.5.	+ 18	05	7.6–8.9	Irregular	—	M
EU Delphini	20	37.9	+ 18	16	5.8–6.9	Irregular	—	M
Beta Doradûs	05	33.6	− 62	29	3.7–4.1	Cepheid	9.8	F–G
R Doradûs	04	36.8	− 62	05	4.8–6.6	Semi-regular	338	M
Z Eridani	02	47.9	− 12	28	7.0–8.6	Semi-regular	80	M
Zeta Geminorum	07	04.1	+ 20	34	3.7–4.1	Cepheid	10.1	F–G
Eta Geminorum	06	14.9	+ 22	30	3.2–3.9	Semi-regular	223	M
R Geminorum	07	07.4	+ 22	42	6.0–14.0	Mira	370	S
U Geminorum	07	55.1	+ 22	00	8.2–14.9	SS Cygni	± 103	M
S Gruis	22	26.1	− 48	26	6.0–15.0	Mira	401	M
Alpha Herculis	17	14.6	+ 14	23	3–4	Semi-regular	± 100	M
S Herculis	16	51.9	+ 14	56	6.4–13.8	Mira	307	M
T Herculis	18	09.1	+ 31	01	6.8–13.9	Mira	165	M
U Herculis	16	25.8	+ 18	54	6.5–13.4	Mira	406	M
AC Herculis	18	30.3	+ 21	52	7.4–9.7	RV Tauri	75	F + K
R Horologii	02	53.9	− 49	53	4.7–14.3	Mira	404	M
R Hydræ	13	29.7	− 23	17	4.0–10.0	Mira	390	M
U Hydræ	10	37.6	− 13	23	4.8–5.8	Semi-regular	450	N
VW Hydri	04	09.1	− 71	18	8.4–14.4	SS Cygni	100	M
S Indi	20	56.4	− 54	19	7.4–14.5	Mira	400	M
T Indi	21	20.2	− 45	01	7.7–9.4	Semi-regular	320	N
R Leonis	09	47.6	+ 11	25	4.4–11.3	Mira	312	M
R Leonis Minoris	09	45.6	+ 34	31	6.3–13.2	Mira	372	M
R Leporis	04	59.6	− 14	48	5.5–11.7	Mira	432	N
Delta Libræ	15	01.1	− 08	31	4.9–5.9	Algol	2.3	B
R Lyncis	07	01.3	+ 55	20	7.2–14.5	Mira	379	S
Beta Lyræ	18	50.1	+ 33	22	3.3–4.3	Beta Lyræ	12.9	B + A
R Lyræ	18	55.3	+ 43	57	3.9–5.0	Semi-regular	46	M
RR Lyræ	19	25.5.	+ 42	47	7.1–8.1	RR Lyræ	0.6	A–F
U Microscopii	20	29.2	− 40	25	7.0–14.4	Mira	334	M
U Monocerotis	07	30.8	− 09	47	6.1–8.1	RV Tauri	92	F–K
V Monocerotis	06	22.7	− 02	12	6.0–13.7	Mira	339	M
R Muscæ	12	42.1	− 69	24	5.9–6.7	Cepheid	7.5	F
R Normæ	15	36.0	− 49	30	6.5–13.9	Mira	493	M
T Normæ	15	44.1	− 54	59	6.2–13.6	Mira	243	M
R Octantis	05	26.1	− 86	23	6.4–13.2	Mira	406	M
RS Ophiuchi	17	50.2	− 06	43	5.3–12.3	Rec. nova	—	M
Alpha Orionis	05	55.2	+ 07	24	0.1–0.9	Semi-regular	± 2110	M
U Orionis	05	55.8	+ 20	10	4.8–12.6	Mira	372	M
W Orionis	05	05.4	+ 01	11	5.9–7.7	Semi-regular	212	N
Kappa Pavonis	18	56.9	− 67	14	3.9–4.7	Type II Cep.	9.1	F
S Pavonis	19	55.2	− 59	12	6.6–10.4	Semi-regular	386	M
Beta Pegasi	23	03.8	+ 28	05	2.3–2.7	Semi-regular	± 38	M
R Pegasi	23	06.6	+ 10	33	6.9–13.8	Mira	378	M
Beta Persei	03	08.2	+ 40	57	2.2–3.4	Algol	2.9	B + G
Rho Persei	03	05.2	+ 38	50	3–4	Semi-regular	± 50	M
X Persei	03	55.4	+ 31	03	6–8	Irregular	—	09.5
Zeta Phœnicis	01	08.4	− 55	15	3.9–4.4	Algol	1.7	B + B

Table A25.2. (*cont.*)

Star	Right ascension		Delination		Range	Type	Period (d)	Spectrum
	hours	minutes	degrees	minutes				
R Phœnicis	23	56.5	−49	47	7.5–14.4	Mira	268	M
R Pictoris	04	46.2	−49	15	6.7–10.0	Semi-regular	164	M
S Pictoris	05	11.0	−48	30	6.5–14.0	Mira	427	M
R Piscium	01	30.6	+02	53	7.1–14.8	Mira	344	M
TX Piscium	23	46.4	+03	29	6.9–7.7	Irregular	—	N
L² Puppis	07	13.5	−44	39	2.6–6.2	Semi-regular	140	M
V Puppis	07	58.2	−49	15	4.7–5.2	Beta Lyræ	1.5	B + B
Z Puppis	07	32.6	−20	40	7.2–14.6	Mira	500	M
T Pyxidis	09	04.7	−32	23	6.3–14.0	Rec. nova	—	Q
R Reticuli	04	33.5	−63	02	6.5–14.0	Mira	278	M
S Sagittæ	19	56.0	+16	38	5.3–6.0	Cepheid	8.4	F–G
U Sagittæ	19	18.8	+19	37	6.6–9.2	Algol	3.4	B + K
WZ Sagittæ	20	07.6	+17	42	7.0–15.5	Rec. nova	—	Q
RR Sagittarii	19	55.9	−29	11	5.6–14.0	Mira	335	M
RY Sagittarii	19	16.5	−33	31	6.0–15	R Coronæ	—	G
R Sagittarii	19	16.7	−19	18	6.7–12.8	Mira	269	M
RR Scorpii	16	55.6	−30	35	5.0–12.4	Mira	279	M
RS Scorpii	16	56.5	−45	06	6.2–13.0	Mira	320	M
RT Scorpii	17	03.5	−36	55	7.0–16.0	Mira	449	M
R Sculptoris	01	27.0	−32	33	5.8–7.7	Semi-regular	370	N
S Sculptoris	00	15.4	−32	03	5.5–13.6	Mira	365	M
R Scuti	18	47.5	−05	42	4.4–8.2	RV Tauri	140	G–K
R Serpentis	15	50.7	+15	08	5.1–14.4.	Mira	356	M
Lambda Tauri	04	00.7	+12	29	3.3–3.8	Algol	3.4	B + A
T Tauri	04	22.0	+19	32	8.4–13.5	T Tauri	—	G–K
R Telescopii	20	14.7	−46	58	7.6–14.8	Mira	462	M
R Trianguli	02	37.0	+34	16	5.4–12.6	Mira	266	M
S Trianguli Aust.	16	01.2	−63	47	6.1–6.8	Cepheid	6.3	F
T Tucanæ	22	40.6	−61	33	7.7–13.8	Mira	251	M
R Ursæ Majoris	10	44.6	+68	47	6.7–13.4	Mira	302	M
T Ursæ Majoris	12	36.4	+59	29	6.6–13.4	Mira	256	M
Z Ursæ Majoris	11	56.5	+57	52	6.8–9.1	Semi-regular	196	M
Z Velorum	09	52.9	−54	11	7.8–14.8	Mira	422	M
AI Velorum	08	14.1	−44	34	6.5–7.1	Delta Scuti	0.11	A–F
R Virginis	12	38.5	+06	59	6.0–12.1	Mira	146	M
S Volantis	07	29.8	73	23	7.7–13.9	Mira	396	M
R Vulpeculæ	21	04.4	+23	49	7.0–14.3	Mira	136	M
T Vulpeculæ	20	51.5	+28	15	5.4–6.1	Cepheid	4.44	F–G
U Vulpeculæ	19	36.6	+20	20	6.8–7.5	Cepheid	7.99	F–G
Z Vulpeculæ	19	21.7	+25	34	7.4–9.2	Algol	2.45	B + A

Table A25.3. *Bright Novæ, 1600–1989 (Magnitude 5 or brighter)*

Nova	Year	Maximum magnitude	Discoverer(s)
CK Vulpeculæ	1670	3	Anthelm
V.841 Ophiuchi	1848	4	Hind
Q Cygni	1876	3	Schmidt
T Aurigæ	1892	4.2	Anderson
V.1059 Sagittarii	1898	4.9	Fleming
GK Persei	1901	0.0	Anderson
DM Geminorum	1903	5.0	Turner
DI Lacertæ	1910	4.6	Espin
DN Geminorum	1912	3.3	Enebo
V.603 Aquilæ	1918	-1.1	Bower
V.476 Cygni	1920	2.0	Denning
RR Pictoris	1925	1.1	Watson
DQ Herculis	1934	1.2	Prentice
CP Lacertæ	1936	1.9	Gomi
V.630 Sagittarii	1936	4.5	Okabayasi
BT Monocerotis	1939	4.3	Whipple and Wachmann
CP Puppis	1942	0.4	Dawson
V.446 Herculis	1960	5.0	Hassell
V.533 Herculis	1963	3.2	Dahlgren and Peltier
HR Delphini	1967	3.7	Alcock
LV Vulpeculæ	1968	4.9	Alcock
FH Serpentis	1970	4.4	Honda
V.1500 Cygni	1975	1.8	Honda

Appendix 26

The Observation of Double Stars

As I have said in the text, measuring the position angles and separations of double stars needs sensitive equipment and endless patience; but the casual observer will find great enjoyment in seeking out the various pairs – at least, I do! Again this is a very incomplete list; I have tried to include doubles of all types, from the very easy to the distinctly difficult.

Of course, the position angles and separations of rapid binaries such as Zeta Herculis and Alpha Centauri alter quite markedly over a period of a very few years. I have given the approximate values as they will be for around 1990.

Star	Right ascension		Declination		Position, angle (degrees)	Separation	Magnitudes	Period (for binary years)
	hours	minutes	degrees	minutes				
Gamma Andromedæ	02	03.9	+42	20	063	9.8	2.3, 4.8	
Gamma² Andromedæ					106	0.5	5.5, 6.3	
Pi Andromedæ	00	36.9	+33	43	173	35.9	4.4, 8.6	
Theta Antliæ	09	44.2	−27	46	005	0.1	5.4, 5.6	
Delta Apodis	16	20.3	−78	42	012	102.9	4.7, 5.1	
Zeta Aquarii	22	28.8	−00	01	200	2.0	4.3, 4.5	856
Gamma Arietis	01	53.5	+19	18	000	7.8	4.8, 4.8	
Theta Aurigæ	05	59.7	+37	13	313	3.6	2.6, 7.1	
Omega Aurigæ	04	59.3	+37	53	359	5.4	5.0, 8.0	
Epsilon Boötis	14	45.0	+27	04	339	2.8	2.5, 4.9	
Zeta Boötis	14	41.1	+13	44	303	1.0	4.5, 4.6	123
Kappa Boötis	14	13.5	+51	47	236	13.4	4.6, 6.6	
Xi Boötis	14	51.4	+19	06	326	7.0	4.7, 7.0	151
Gamma Cæli	05	04.4	−35	29	308	2.9	4.6, 8.1	
Zeta Cancri	08	12.2	+17	39	AB+C 088	5.7	5.0, 6.2	1150
					AB 182	0.6	5.3, 6.0	60
					AB+D 108	278.9	9.7	
Alpha Canum Venaticorum	12	56.0	+38	19	229	19.4	2.9, 5.5	
Alpha Canis Majoris	06	45.1	−16	43	005	4.5	−1.5, 8.5	50
Epsilon Canis Majoris	06	58.6	−28	58	161	7.5	1.5, 7.4	
Alpha Capricorni	20	18.1	−12	33	291	377.7	3.6, 4.2	
Upsilon Carinæ	09	47.1	−65	04	127	5.0	3.1, 61	
Eta Cassiopeiæ	00	49.1	+57	49	293	12.2	3.4, 7.5	480
Iota Cassiopeiæ	02	29.1	+67	24	232	2.4	4.6, 6.9	840
Lambda Cassiopeiæ	00	31.8	+54	31	176	0.5	5.3, 5.6	

Alpha Centauri	14	39.6	−60	50	215	19.7	0.0, 1.2	80
Gamma Centauri	12	41.5	−48	58	353	1.4	2.9, 2.9	85
Beta Cephei	21	28.7	+70	34	249	13.3	3.2, 7.9	
Kappa Cephei	20	08.9	+72	43	122	7.4	4.4,8.4	
Xi Cephei	22	03.8	+64	38	277	7.7	4.4, 6.5	
Gamma Ceti	02	43.3	+03	14	294	2.8	3.5, 7.3	
Omicron Ceti	02	19.3	−02	59	085	0.3	Variable, 12.0	400
66 Ceti	02	12.8	−02	24	234	16.5	5.7, 7.5	
Mu Crucis	12	54.6	−57	11	017	34.9	4.0, 5.2	
Delta Chamæleontis	10	45.3	−80	28	076	0.6	6.1, 6.4	
Epsilon Chamæleontis	11	59.6	−78	13	188	0.9	5.4, 6.0	
Gamma Circini	15	23.4	−59	19	033	0.6	5.1, 5.5	180
Gamma Coronæ Australis	19	06.4	−37	04	109	1.3	4.8, 5.1	120
Kappa Coronæ Australis	18	33.4	−38	44	369	21.6	5.9, 5.9	
Zeta Coronæ Borealis	15	39.4	+36	38	305	6.3	5.1, 6.0	
Eta Coronæ Borealis	15	23.2	+30	17	030	1.0	5.6,5.9	42
Sigma Coronæ Borealis	16	14.7	+33	52	234	7.0	5.6, 6.6	1000
Delta Corvi	12	29.9	−16	31	214	24.2	3.0,9.2	
Alpha Crucis	12	26.6	−63	06	AB 115	4.4	1.4,1.9	
					AC 202	90.1	1.0, 4.9	
Gamma Crucis	12	31.2	−57	07	AB 031	110.6	1.6, 6.7	
					AC 082	155.2	9.5	
Beta Cygni	19	30.7	+27	58	054	34.4	3.1, 5.1	
Delta Cygni	19	45.0	+45	07	225	2.4	2.9, 6.3	
61 Cygni	21	07	+38	45	140	27.0	5.2, 6.0	
Gamma Delphini	20	46.7	+16	07	268	9.6	4.5, 5.5	
Mu Draconis	17	05.3	+54	28	020	1.9	5.7, 5.7	482
Nu Draconis	17	32.2	+55	11	312	61.9	4.9, 4.9	
Delta Equulei	21	14.5	+10	00	030	0.3	5.2, 5.3	5.7

Star	Right ascension		Declination		Position, angle (degrees)	Separation	Magnitudes	Period (for binary years)
	hours	minutes	degrees	minutes				
Theta Eridani	02	58.3	− 40	18	088	8.2	3.4, 4.5	
Omicron² Eridani	04	15.2	− 07	39	104	83.4	4.4, 9.5	
Alpha Fornacis	03	12.1	− 28	59	298	4.0	4.0, 7.0	314
Omega Fornacis	02	33.8	− 28	14	244	10.8	5.0, 7.7	
Alpha Geminorum	07	34.6	+ 31	53	AB 088	2.5	1.9, 2.9	420
					AC 164	72.5	1.6, 8.8	
Delta Geminorum	07	20.1	+ 21	59	223	6.0	3.5, 8.2	1200
Eta Geminorum	06	14.9	+ 22	30	266	1.4	Variable, 8.8	474
Kappa Geminorum	07	44.4	+ 24	34	240	7.1	3.6, 8.1	
Theta Gruis	23	06.9	− 43	31	075	1.1	4.5, 7.0	
Alpha Herculis	17	14.6	+ 14	23	107	4.7	Variable, 5.4	3600
Zeta Herculis	16	41.3	+ 31	36	089	1.6	2.9, 5.5	34.5
Kappa Herculis	17	14.6	+ 17	03	012	28.4	5.3, 6.5	
Beta Hydrae	11	52.9	− 33	54	008	0.9	4.7, 5.5	
Epsilon Hydrae	08	46.8	+ 06	25	AB 295	0.2	3.8, 4.7	890
					AB+C 281	2.8	6.8	
Theta Hydrae	21	19.9	− 53	27	275	6.0	4.5, 7.0	
Theta Indi	09	14.4	+ 02	19	197	29.4	3.9, 9.9	
Alpha Leonis	10	08.4	+ 11	58	307	176.9	1.4, 7.7	
Gamma Leonis	10	20.0	+ 19	51	124	4.3	2.2, 3.5	619
Omega Leonis	09	28.5	+ 09	03	053	0.5	5.9, 6.5	118
Beta Leonis Minoris	10	27.9	+ 36	42	250	0.2	4.4, 6.1	37
Beta Leporis	05	28.2	− 20	46	330	2.5	2.8, 7.3	
Gamma Leporis	05	44.5	− 22	27	350	96.3	3.7, 6.3	
Kappa Leporis	05	13.2	− 12	56	358	2.6	4.5, 7.4	

Name								
Alpha Libræ	14	50.9	−16	02	314	231.0	2.8, 5.2	
Epsilon Lupi	15	22.7	−44	41	247	0.6	3.7, 7.2	
Xi Lupi	15	56.9	−33	58	049	10.4	5.3, 5.8	
Epsilon Lyræ	18	44.3	+39	40	AB+CD 173	207.7	4.7, 5.1	
					AB 357	2.6	5.0, 6.1	
					CD 094	2.3	5.2, 5.5	
Zeta Lyræ	18	44.8	+37	36	150	43.7	4.3, 5.9	
Theta² Microscopii	21	24.4	−41	00	267	0.5	6.4, 7.0	
Beta Monocerotis	06	28.8	−07	02	AB 132	7.3	4.7, 5.2	
					AC 124	10.0	6.1	
Epsilon Monocerotis	06	23.8	+04	36	027	13.4	4.5, 6.5	
Beta Muscæ	12	46.3	−68	06	014	1.4	3.7, 4.0	
Epsilon Normæ	16	27.2	−47	33	335	22.8	4.8, 7.5	
Iota Octantis	12	55.0	−85	07	230	0.6	6.0, 6.5	
Lambda Octantis	21	50.9	−82	43	070	3.1	5.4, 7.7	
Eta Ophiuchi	17	10.4	−15	43	247	0.5	3.0, 3.5	84
Rho Ophiuchi	16	25.6	−23	27	344	3.1	5.3, 6.0	
Beta Orionis	05	14.5	−08	12	202	9.5	0.1, 6.8	
Delta Orionis	05	32.0	−00	18	359	52.6	2.2, 6.3	
Zeta Orionis	05	40.8	−01	57	AB 162	2.4	1.9, 4.0	1509
					AC 010	57.6	9.9	
Eta Orionis	05	24.5	−02	24	AB 080	1.5	3.8, 4.8	
					AC 051	115.1	9.4	
Theta Orionis	05	35.3	−05	23	AB 031	8.8	6.7, 7.9	
					AC 132	12.8	5.1	
					AD 096	21.5	6.7	
Iota Orionis	05	35.4	−05	55	141	11.3	2.8, 6.9	
Eta Pegasi	22	43.0	+30	13	339	90.4	2.9, 9.9	
Beta Phœnicis	01	06.1	−46	43	346	1.4	4.0, 4.2	

Star	Right ascension		Declination		Position, angle (degrees)	Separation	Magnitudes	Period (for binary years)
	hours	minutes	degrees	minutes				
Iota Pictoris	04	50.9	− 53	28	058	12.3	5.6, 6.4	
Xi Pavonis	18	23.2	− 61	30	154	3.3	4.4, 8.6	
Beta Piscis Australis	22	31.5	− 32	21	172	30.3	4.4, 7.9	
Alpha Piscium	02	02.0	+ 02	46	279	1.9	4.2, 5.1	933
Zeta Piscium	01	13.7	+ 07	35	063	23.0	5.6, 6.5	
Psi Piscium	01	05.6	+ 21	28	159	30.0	5.6, 5.8	
Zeta Sagittae	19	49.0	+ 19	09	AB+C 311	8.6	5.5, 8.7	23
					AB 163	0.3	5.5, 6.2	
Zeta Sagittarii	19	02.6	− 29	53	320	0.3	3.3, 3.4	21
Eta Sagittarii	18	17.6	− 36	46	105	3.6	3.2, 7.8	
Kappa² Sagittarii	20	23.9	− 42	25	234	0.8	6.9, 6.9	
Pi Sagittarii	19	09.8	− 21	01	AB 150	0.1	3.7, 3.7	
					AB+C 122	0.4	5.9	
Alpha Scorpii	16	29.4	− 26	26	273	2.7	1.2, 5.4	878
Beta Scorpii	16	05.4	− 19	48	021	13.6	2.6, 4.9	
Nu Scorpii	16	12.0	− 19	28	AB 003	0.9	4.3, 6.8	46
					AC 337	41.1	6.4	
Delta Serpentis	15	34.8	+ 10	32	177	4.4	4.1, 5.2	3168
Theta Serpentis	18	56.2	+ 04	12	104	22.3	4.5, 4.5	
Kappa-67 Tauri	04	25.4	+ 22	18	173	339	4.2, 5.3	
Theta Tauri	04	28.7	+ 15	32	346	337.4	3.4, 3.8	
Sigma Tauri	04	39.3	+ 15	55	193	431.2	4.7, 5.1	
Iota Trianguli	02	15.9	+ 33	21	240	2.3	5.4, 7.0	
Beta Tucanae	00	31.5	− 62	58	169	27.1	4.4, 4.8	
Beta² Tucanae	00	31.6	− 62	58	295	0.6	4.8, 6.0	44

Alpha Ursæ Majoris	11	03.7	+ 61	45	283	0.7	1.9, 4.8	45
Zeta Ursæ Majoris	13	23.9	+ 54	56	AB 152	14.4	2.3, 4.0	Mizar–Alcor
					AC 071	708.7	2.1, 4.0	
Kappa Ursæ Majoris	09	03.6	+ 47	09	258	0.1	4.2, 4.4	70
Xi Ursæ Majoris	11	18.2	+ 31	32	060	1.3	4.8, 4.8	60
Alpha Ursæ Minoris	02	31.8	+ 89	16	218	18.4	2.0, 9.0	
Gamma Velorum	08	09.5	− 47	20	220	41.2	1.9, 4.2	
Delta Velorum	08	44.7	− 54	43	153	2.6	2.1, 5.1	
Mu Velorum	10	46.8	− 49	25	055	2.3	2.7, 6.4	
Gamma Virginis	12	41.7	− 01	27	287	3.0	3.5, 3.5	116
Gamma² Volantis	07	08.8	− 70	30	300	13.6	4.0, 5.9	171
Alpha-8 Vulpeculæ	19	28.7	+ 24	40	028	413.7	4.4, 5.8	

Appendix 27

The Observation of Star-Clusters and Nebulæ

There are many amateurs who make a point of observing all the Messier objects — which is not difficult, because even the faintest of them, M.76 (the Little Dumbbell, a planetary in Perseus) is no dimmer than magnitude 11.5. I have also listed some of the interesting objects not included by Messier.

Apart from sheer interest, there is valuable work to be done in hunting for supernovæ in external galaxies; and as I have pointed out in the text, some amateurs have had remarkable success in recent years.

Table A27.1. *Messier's catalogue*

M	NGC	Right ascension		Declination		Constellation	Magnitude	Type
		hours	minutes	degrees	minutes			
1	1952	05	34.5	+ 22	01	Taurus	8.4	Supernova remnant
2	7089	21	35.5	− 00	49	Aquarius	6.5	Globular
3	5272	13	42.2	+ 28	23	Canes Venatici	6.4	Globular
4	6121	16	23.6	− 26	32	Scorpius	5.9	Globular
5	5904	15	18.6	+ 02	05	Serpens	5.8	Globular
6	6405	17	40.1	− 32	13	Scorpius	4.2	Open cluster
7	6475	17	53.9	− 34	49	Scorpius	3.3	Open cluster
8	6523	18	03.8	− 24	23	Sagittarius	5.8	Nebula (Lagoon)
9	6333	17	19.2	− 18	31	Ophiuchus	7.9	Globular
10	6524	16	57.1	− 04	06	Ophiuchus	6.6	Globular
11	6705	18	51.1	− 06	16	Scutum	5.8	Open cluster (Wild Duck)
12	6218	16	47.2	− 01	57	Ophiuchus	6.6	Globular
13	6205	16	41.7	+ 36	28	Hercules	5.9	Globular (Hercules Cluster)
14	6402	17	37.6	− 03	15	Ophiuchus	7.6	Globular
15	7078	21	30.0	+ 12	10	Pegasus	6.4	Globular
16	6611	18	18.8	− 13	47	Serpens	6.0	Nebula + cluster (Eagle Nebula)
17	6618	18	20.8	− 16	11	Sagittarius	7.0	Nebula (Omega)
18	6613	18	19.9	− 17	08	Sagittarius	6.9	Open cluster
19	6273	18	02.6	− 26	16	Ophiuchus	7.2	Globular
20	6514	18	02.6	− 23	02	Sagittarius	8.5	Nebula (Trifid)
21	6531	18	04.6	− 22	30	Sagittarius	5.9	Open cluster
22	6656	18	36.4	− 23	54	Sagittarius	5.1	Globular

Table A27.1. (*cont.*)

M	NGC	Right ascension hours	minutes	Declination degrees	minutes	Constellation	Magnitude	Type
23	6494	17	56.8	− 19	01	Sagittarius	5.5	Open cluster
24	—	18	16.9	− 18	29	Sagittarius	4.5	Star-cloud in the Milky Way
25	IC 4725	18	31.6	− 19	15	Sagittarius	4.6	Open cluster
26	6694	18	45.2	− 09	24	Scutum	8.0	Open cluster
27	6853	19	59.6	+ 22	43	Vulpecula	8.1	Planetary (Dumbbell)
28	6626	18	24.5	− 24	52	Sagittarius	6.9	Globular
29	6913	20	23.9	+ 38	32	Cygnus	6.6	Open cluster
30	7099	21	40.4	− 23	11	Capricornus	7.5	Globular
31	224	00	42.7	+ 41	16	Andromeda	3.4	Spiral galaxy (Great Spiral)
32	221	00	42.7	+ 40	52	Andromeda	8.2	Elliptical galaxy
33	598	01	33.9	+ 30	39	Triangulum	5.7	Spiral galaxy
34	1039	02	42.0	+ 42	47	Perseus	5.2	Open cluster
35	2168	06	08.9	+ 24	20	Gemini	5.1	Open cluster
36	1960	05	36.1	+ 34	08	Auriga	6.0	Open cluster
37	2099	05	52.4	+ 32	33	Auriga	5.6	Open cluster
38	1912	05	28.7	+ 35	50	Auriga	6.4	Open cluster
39	7092	21	32.2	+ 48	26	Cygnus	4.6	Open cluster
40	—	12	22.4	+ 58	05	Ursa Major	9	Double star (Winnecke 4)
41	2287	06	47.0	− 20	44	Canis Major	4.5	Open cluster
42	1976	05	35.4	− 05	27	Orion	4	Nebula (Great Nebula in Orion)
43	1982	05	35.6	− 05	16	Orion	9	Nebula (Part of M.42)
44	2632	08	40.1	+ 19	59	Cancer	3.3	Open cluster (Præsepe)
45	—	03	47.0	+ 24	07	Taurus	1.2	Open cluster (Pleiades)
46	2437	07	41.8	− 14	49	Puppis	6.7	Open cluster
47	2422	07	36.6	− 14	30	Puppis	4.4	Open cluster
48	2548	08	13.8	− 05	48	Hydra	5.8	Open cluster
49	4472	12	29.8	+ 08	00	Virgo	8.4	Elliptical galaxy
50	2323	07	03.2	− 08	20	Monoceros	5.9	Open cluster
51	5194–5	13	29.9	+ 47	12	Canes Venatici	8.1	Spiral galaxy (Whirlpool)
52	7654	23	24.2	+ 61	35	Cassiopeia	6.9	Open cluster
53	5024	13	12.9	+ 18	10	Coma Berenices	7.7	Globular
54	6715	18	55.1	− 30	29	Sagittarius	7.7	Globular
55	6809	19	40.0	− 30	58	Sagittarius	7.0	Globular
56	6779	19	16.6	+ 30	11	Lyra	8.2	Globular
57	6720	18	53.6	+ 33	02	Lyra	9.0	Planetary (Ring)
58	4579	12	37.7	+ 11	49	Virgo	9.8	Spiral galaxy
59	4621	12	42.0	+ 11	39	Virgo	9.8	Elliptical galaxy
60	4649	12	43.7	+ 11	33	Virgo	8.8	Elliptical galaxy
61	4303	12	21.9	+ 04	28	Virgo	9.7	Spiral galaxy
62	6266	17	01.2	− 30	07	Ophiuchus	6.6	Globular
63	5055	13	15.8	+ 42	02	Canes Venatici	8.6	Spiral galaxy
64	4826	12	56.7	+ 21	41	Coma Berenices	8.5	Spiral galaxy (Black-eye)
65	3623	11	18.9	+ 13	05	Leo	9.3	Spiral galaxy
66	3627	11	20.2	+ 12	59	Leo	9.0	Spiral galaxy

Table A27.1. (*cont.*)

M	NGC	Right ascension		Declination		Constellation	Magnitude	Type
		hours	minutes	degrees	minutes			
67	2682	08	50.4	+ 11	49	Cancer	6.9	Open cluster
68	4590	12	39.5	− 26	45	Hydra	8.2	Globular
69	6637	18	31.4	− 32	21	Sagittarius	7.7	Globular
70	6681	18	43.2	− 32	18	Sagittarius	8.1	Globular
71	6838	19	53.8	+ 18	47	Sagitta	8.3	Globular
72	6981	20	53.5	− 12	32	Aquarius	9.4	Globular
73	6994	20	58.9	− 12	38	Aquarius	9	Four faint stars
74	628	01	36.7	+ 15	47	Pisces	9.2	Spiral galaxy
75	6864	20	06.1	− 21	55	Sagittarius	8.6	Globular
76	650–1	01	42.4	+ 51	34	Perseus	11.5	Planetary (Little Dumbbell)
77	1068	02	42.7	− 00	01	Cetus	8.8	Spiral galaxy
78	2068	05	46.7	+ 00	03	Orion	8	Nebula
79	1904	05	24.5	− 24	33	Lepus	8.0	Globular
80	6093	16	17.0	− 22	59	Scorpius	7.2	Globular
81	3031	09	55.6	+ 69	04	Ursa Major	6.8	Spiral galaxy
82	3034	09	55.8	+ 69	41	Ursa Major	8.4	Irregular galaxy
83	5236	13	37.0	− 29	52	Hydra	7.6	Spiral galaxy
84	4374	12	25.1	+ 12	53	Virgo	9.3	Spiral galaxy
85	4382	12	25.4	+ 18	11	Coma Berencies	9.2	Spiral galaxy
86	4406	12	26.2	+ 12	57	Virgo	9.2	Elliptical galaxy
87	4487	12	30.8	+ 12	24	Virgo	8.6	Elliptical galaxy (Virgo A)
88	4501	12	32.0	+ 14	25	Coma Berenices	9.5	Spiral galaxy
89	4552	12	35.7	+ 12	33	Virgo	9.8	Elliptical galaxy
90	4569	12	36.8	+ 13	10	Virgo	9.5	Spiral galaxy
91	—	—		—		Coma Berenices	Missing	
92	6341	17	17.1	+ 43	08	Hercules	6.5	Globular
93	2447	07	44.6	− 23	52	Puppis	6.2	Open cluster
94	4736	12	50.9	+ 41	07	Canes Venatici	8.1	Spiral galaxy
95	3351	10	44.0	+ 11	42	Leo	9.7	Barred spiral galaxy
96	3368	10	46.8	+ 11	49	Leo	9.2	Spiral galaxy
97	3587	11	14.8	+ 55	01	Ursa Major	11.2	Planetary (Owl)
98	4192	12	13.8	+ 14	25	Coma Berenices	10.1	Spiral galaxy
99	4254	12	18.8	+ 14	25	Coma Berenices	9.8	Spiral galaxy
100	4321	12	22.9	+ 15	49	Coma Berenices	9.4	Spiral galaxy
101	5457	14	03.2	+ 54	21	Ursa Major	7.7	Spiral galaxy
102	—	—		—		Ursa Major	Missing	
103	581	01	33.2	+ 60	42	Cassiopeia	7.4	Open cluster
104	4594	12	40.0	− 11	37	Virgo	8.3	Spiral galaxy
105	3379	10	47.8	+ 12	35	Leo	9.3	Elliptical galaxy
106	4258	12	19.0	+ 47	18	Canes Venatici	8.3	Spiral galaxy
107	6171	16	32.5	− 13	03	Ophiuchus	8.1	Globular
108	3556	11	11.5	+ 55	40	Ursa Major	9.8	Spiral galaxy
109	3992	11	57.6	+ 53	23	Andromeda	9.8	Spiral galaxy

Note: Messier's original catalogue ended with M.103. Various astronomers have since added the rest − even to M.110, better known as NGC 205, one of the companions of the Great Spiral in Andromeda, M.31 (the other companion is M.32). Several Messier numbers were 'missing'. M.47 is now generally identified as the bright Puppis cluster. M.24 is merely a brightening of the Milky Way. M.48 is generally identified with the Hydra cluster. M.91 has been identified with the 10th-magnitude spiral NGC 4548, but not with certainty; it might possibly have been a comet. According to Messier's colleague Pierre Méchain, M.102 was an accidental re-observation of M.101.

Table A27.2. *Interesting clusters and nebulae not in Messier's catalogue*

Object	Right ascension		Declination			Constellation	Magnitude	Type
	hours	minutes	degrees	minutes				
NGC 205	00	40.4	+41	41		Andromeda	8.0	Elliptical galaxy
NGC 752	01	57.8	+37	41		Andromeda	5.7	Open cluster
NGC 7009	21	04.2	−11	22		Aquarius	8.3	Planetary (Saturn Nebula)
NGC 7293	22	29.6	−20	48		Aquarius	6.5	Planetary (Helix Nebula)
NGC 6709	18	51.5	+10	21		Aquila	6.7	Open cluster
NGC 6755	19	07.8	+04	14		Aquila	7.5	Open cluster
NGC 6790	19	23.2	+01	31		Aquila	10.2	Planetary
NGC 6352	17	25.5	−48	25		Ara	8.1	Globular
NGC 6362	17	31.9	−67	03		Ara	8.3	Globular
NGC 6397	17	40.7	−53	40		Ara	5.6	Globular
NGC 1857	05	20.2	+39	21		Auriga	7.0	Open cluster
IC 342	03	46.8	+68	06		Camelopardalis	9.2	Barred spiral galaxy
NGC 1502	04	07.7	+62	20		Camelopardalis	5.7	Open cluster
NGC 2360	07	17.8	−15	37		Canis Major	7.2	Open cluster
NGC 3372	10	43.8	−59	52		Carina	6.2	Nebula (Eta Carinæ)
NGC 2516	07	58.3	−60	52		Carina	3.8	Open cluster
NGC 3532	11	06.4	−58	40		Carina	3.0	Open cluster
IC 2581	10	27.4	−57	38		Carina	4.3	Open cluster
IC 2602	10	43.2	−64	24		Carina	1.9	Open Cluster (Theta Carinæ)
NGC 7789	23	57.0	+56	44		Cassiopeia	6.7	Open cluster
NGC 457	01	19.1	+58	20		Cassiopeia	6.4	Open cluster (Phi Cassiopeiæ)
NGC 7635	23	20.7	+61	12		Cassiopeia	7	Nebula (Bubble)
NGC 3766	11	35.1	−61	37		Centaurus	5.3	Open cluster
IC 2944	11	36.6	−63	02		Centaurus	4.5	Open cluster (Lambda Centauri)

Table A27.2. (*cont.*)

Object	Right ascension		Declination		Constellation	Magnitude	Type
	hours	minutes	degrees	minutes			
NGC 5139	13	26.8	−47	29	Centaurus	3.6	Globular (Omega Centauri)
NGC 5367	13	57.7	−39	59	Centaurus	4.3	Nebula
NGC 5128	13	25.5	−43	01	Centaurus	7.0	Galaxy (Centaurus A)
NGC 5617	14	29.8	−60	43	Centaurus	6.3	Open cluster
NGC 5662	14	35.2	−56	33	Centaurus	5.5	Open cluster
NGC 5316	13	53.9	−61	52	Centaurus	6.0	Open cluster
IC 1396	21	39.1	+57	30	Cepheus	3.5	Open cluster
Sh2-155	22	56.8	+62	37	Cepheus	7.7	Nebula (Cave Nebula)
NGC 247	00	47.1	−20	46	Cetus	8.9	Spiral galaxy
NGC 5823	15	05.7	−55	36	Circinus	7.9	Open cluster
NGC 1851	05	14.1	−40	03	Columba	7.3	Globular
Mel 111	12	25	+26		Coma Berenices	4	Open cluster (Coma Berenices)
NGC 4103	12	06.7	−61	15	Crux	7.4	Open cluster
NGC 4755	12 –	53.6	−60	20	Crux	4.2	Open cluster; Kappa Crucis (Jewel Box)
—	12	53	−63		Crux	—	Dark Nebula (Coal Sack)
NGC 6826	19	44.8	+50	31	Cygnus	9.8	Planetary (Blinking Nebula)
NGC 6811	19	38.2	+46	34	Cygnus	6.8	Open cluster
NGC 6960	20	45.7	+30	43	Cygnus	—	Nebula (Filamentary)
NGC 6992/5	20	56.4	+31	43	Cygnus	—	Supernova remnant (Veil Nebula)
NGC 7000	20	58.8	+44	20	Cygnus	6	Nebula (North America Nebula)
—	05	23.6	−69	45	Dorado	1	Galaxy (Large Cloud of Magellan)

NGC	h	m	°	′	Constellation	Mag.	Description
NGC 2070	05	38.7	−69	06	Dorado	—	Nebula (Tarantula, 30 Doradûs, in LMC)
NGC 2392	07	29.2	+20	55	Gemini	10.5	Planetary (Eskimo Nebula)
NGC 6210	16	44.5	+23	49	Hercules	9.3	Planetary
NGC 1261	03	12.3	−55	13	Horologium	8.4	Globular
NGC 3242	10	24.8	−18	38	Hydra	8.6	Planetary (Ghost of Jupiter)
NGC 7243	22	15.3	+49	53	Lacerta	6.4	Open cluster
NGC 5822	15	05.2	−54	21	Lupus	6.5	Open cluster
NGC 2237–9	06	32.3	+05	03	Monoceros	—	Nebula (Rosette)
NGC 2244	06	32.4	+04	52	Monoceros	4.8	Open cluster (in Rosette Nebula)
NGC 2264	06	40.9	+09	54	Monoceros	Variable	Nebula (Cone Nebula, S Monocerotis)
NGC 2261	06	39.2	+08	44	Monoceros	Variable	Nebula (Hubble's Variable Nebula; R Monocerotis)
NGC 4465	12	30.0	−64	48	Musca	7.2	Open cluster
NGC 4372	12	25.8	−72	40	Musca	7.8	Globular
NGC 4833	12	59.6	−70	53	Musca	7.3	Globular
NGC 6087	16	18.9	−57	54	Norma	5.4	Open cluster (S Normæ)
NGC 2175	06	09.8	+20	19	Orion	6.8	Open cluster
NGC 6752	19	10.9	−59	59	Pavo	5.4	Globular
NGC 869/884	02	21	+57	08	Perseus	4	Double cluster; Sword-Handle (Chi-h Persei)
NGC 1499	04	00.7	+36	37	Perseus	4	Nebula (California Nebula)
NGC 2451	-07	45.4	−37	58	Puppis	2.8	Open cluster
NGC 6302	17	13.7	−37	06	Scorpius	12.8	Nebula (Bug Nebula)
NGC 55	00	14.9	−39	11	Sculptor	8.2	Barred spiral galaxy
NGC 253	00	47.6	−25	17	Sculptor	7.1	Spiral galaxy
—	04	27	+16		Taurus	1	Open cluster (Hyades)

Table A27.2. (cont.)

Object	Right ascension		Declination		Constellation	Magnitude	Type
	hours	minutes	degrees	minutes			
NGC 1647	04	46.0	+19	04	Taurus	6.4	Open cluster
NGC 6025	16	03.7	−60	30	Triangulum Australe	5.2	Open cluster
NGC 104	00	24.1	−72	05	Tucana	4.0	Globular (47 Tucanæ)
NGC 362	01	03.2	−70	51	Tucana	6.6	Globular
—	00	52.7	−72	50	Tucana	2.3	Galaxy (Small Cloud of Magellan)
NGC 2547	08	10.7	−49	16	Vela	4.7	Open cluster
IC 2391	08	40.2	−53	04	Vela	2.5	Open cluster (o Velorum)
NGC 3201	10	17.6	−46	25	Vela	6.7	Globular
NGC 3132	10	07.7	−40	26	Vela	8.2	Planetary
NGC 6940	20	34.6	+28	18	Vulpecula	6.3	Open cluster

Appendix 28

The Star Maps

When I set out to learn the star-patterns, which was when I had reached the advanced age of seven, I made a pious resolve to identify one new constellation on every clear night. The method I used was to begin with a few unmistakable groups, and use these as pointers to the rest. Starting with Ursa Major and Orion, it is possible to learn how to recognize all the constellations visible from Britain; when I first went to the southern hemisphere I could still use Orion, plus the Southern Cross.

I have therefore given a few key maps, and then some more detailed charts. These are certainly not precision charts, and are not intended to be, but I hope they will serve as a guide. In the constellation notes, all stars down to magnitude 3.5 have been listed under the heading 'Chief Stars'; I have referred to some interesting objects, of which more details will be found in the lists in 25–27. Not all the objects in these lists are shown on the maps, because on this scale it would have made the maps themselves too crowded, but they can be looked up on a more elaborate chart.

Planets, naturally, cannot be shown, because they move around; but their positions for any time can be looked up in the monthly charts in *Sky and Telescope* or *Astronomy Now*.

THE NORTHERN-HEMISPHERE VIEW

Map I. Key Map: Ursa Major (the Great Bear).

Almost everyone must know the Great Bear, with its seven famous stars making up the pattern which is often nicknamed the Plough or the Big Dipper. It is so far north in the sky that it never sets over the British Isles or the northern United States.

The seven Plough stars are: Eta (Alkaid), Zeta (Mizar), Epsilon (Alioth), Delta (Megrez), Gamma (Phad), Beta (Merak) and Alpha (Dubhe). All are of around the second magnitude apart from Megrez, which is below the third. The proper names are frequently used; in addition, Merak and Dubhe are known as 'the Pointers'.

The first step after having identified the Bear is to locate the Pole Star. Imagine a line drawn from Merak through Dubhe, and prolonged; it will reach a second-magnitude

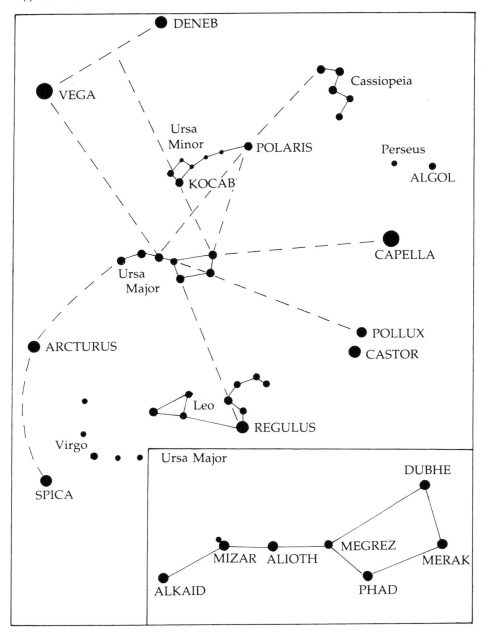

Map I. Key map: Ursa Major (the Great Bear).

star rather 'out on its own', and this is Polaris. The Little Bear, Ursa Minor, can then be picked out, curving back toward the Great Bear itself. The stars of Ursa Minor are much fainter, but one of them, the orange Kocab, is of the second magnitude.

Now take a line from Alioth through Polaris. Prolonged for an equal distance on the far side of Polaris, it will reach five brightish stars (magnitudes 2–3) arranged in a rough W form. This is Cassiopeia, which, like the Bears, never sets over Britain. When Cassiopeia is high up, the Great Bear is low down, and vice versa.

A line from Megrez through Dubhe will come eventually to Capella, which is one of the brightest stars in the entire sky. It is circumpolar from England, but at its lowest, as during summer evenings, it almost touches the horizon, and from most of the United States it actually sets. In winter evenings it is high up, and may indeed pass overhead. If you see a really bright star straight above you, it can only be Capella or Vega; Capella is yellowish, and has a small triangle of stars close beside it, whereas Vega is steely blue. Vega can be found from a line beginning at Phad, passing between Megrez and Alioth, and prolonged for some distance across the sky.

The remaining stars shown in Map I are not circumpolar from Britain. The Twins, Castor and Pollux, can be found by means of a line from Megrez through Merak; they are at their best in winter. Regulus and the other stars of Leo (the Lion), found by a line from Megrez through Phad, seem to follow the Twins in the sky; the curved arrangement of stars rather like the mirror image of a question-mark, of which Regulus is the brightest member, is known as the Sickle of Leo, and is easy to recognize. Even easier is Arcturus, a lovely orange star slightly brighter than Capella or Vega; it can be found by means of a line from Mizar through Alkaid, curved to follow the curve of the 'Bear's tail'. If the line is continued past Arcturus it will lead to another first-magnitude star, Spica in Virgo. Arcturus and Spica are prominent from northern latitudes all through spring and early summer evenings.

Map II. Key Map: Orion.

It is a pity that Orion is not circumpolar in Britain, as it is a magnificent 'signpost' as well as being a beautiful constellation in itself, but at least it is on view from Britain and the United States for a reasonable part of the year, and because the celestial equator passes through it Orion can be seen from every inhabited country. The periods of visibility of Orion from latitudes such as those of England can be judged from the following:

1 January: Rises 4 p.m., highest 10 p.m., sets 5 a.m.
1 April: Rises in daylight, highest in daylight, sets 11 p.m.
1 July: Rises 4 a.m., highest in daylight, sets in daylight.
1 October: Rises 10 p.m., highest 5 a.m., sets in daylight.

Of course these times are very rough, because Orion covers a considerable area and takes hours to 'rise' or 'set', but from these notes you can see that it is at its best during winter and early autumn. I have given the times in GMT.

Orion has a very distinctive pattern, and its stars are brighter than those of Ursa

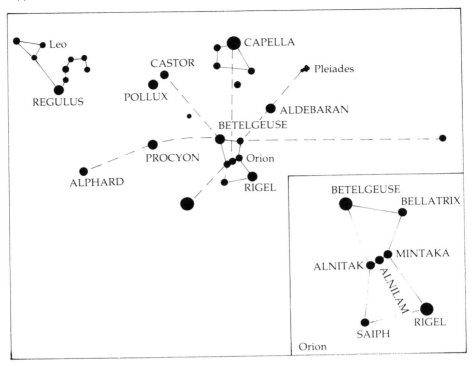

Map II. Key map: Orion.

Major. Two – the orange Betelgeuse and the white Rigel – are of the first magnitude; Rigel is only marginally fainter than Vega or Capella. The other stars in the main pattern are of the second magnitude. Mintaka, Alnilam and Alnitak form the famous Belt; the celestial equator passes very close to Mintaka.

The first-magnitude stars in the key map are easy to find if Orion is on view. The Belt stars point downward to Sirius, the brightest star in the sky, and upward to Aldebaran in Taurus, an orange-red star of the first magnitude. Beyond Aldebaran lies the famous star-cluster of the Pleiades or Seven Sisters.

Bellatrix and Betelgeuse point more or less to Procyon, in Canis Minor (the Little Dog); Procyon is not much fainter than Rigel. If this line is continued and curved slightly, it will reach a reddish second-magnitude star, Alphard in Hydra, which is nicknamed 'the Solitary One' because it lies in a very barren region. The twins, Castor and Pollux, can be found from a line from Rigel through Betelgeuse; since they can also be found from Ursa Major, this links the two key maps. Capella can be located by a line from Saiph through Alnitak. Diphda in Cetus, the other star shown in the key map, is less easy to identify. It is only of the second magnitude, and is frequently visible when Orion is below the horizon.

Undoubtedly a winter evening is the best time to begin star recognition, because both our 'signposts', Orion and Ursa Major, can be seen. If you start in summer, you must do without Orion, but on its own the Great Bear can provide guides to most of the main groups.

Each of the charts given in the next section contains at least one key map object. I think you will find that even though the stars seem at first sight to be arranged in a chaotic manner, it takes surprisingly little time to find one's way about.

Map III. Ursa Major, Ursa Minor, Draco, Cepheus, Camelopardalis.

This is the North Polar region, circumpolar in Britain and the northern United States. Of course the Pole Star cannot be seen from the southern hemisphere; Ursa Major is accessible in part from most of South Africa and Australia, but not from most of New Zealand.

Ursa Major.   The chief stars are Epsilon (Alioth) and Alpha (Dubhe) (each 1.8), Eta (Alkaid) 1.9, Zeta (Mizar) 2.1, Beta (Merak) and Gamma (Phad) each 2.4, Psi and Mu (each 3.0), Iota (3.1), Theta (3.2), Delta (Megrez) 3.3, Omicron and Lambda (each 3.4) and Nu (3.5). Part of the constellation extends on to Map VI.

Doubles. Mizar is the most celebrated double in the sky. It forms a naked-eye pair with Alcor (4.0), and a small telescope will separate the two components of Mizar itself, as well as showing 'Ludwig's Star' in the same field. Kappa and Xi are both close pairs with nearly equal components.

Variables. The Mira stars R and T are both fairly bright when at maximum, though at minimum they require apertures of fair size. The red semi-regular Z is easy to find, and is a good binocular object; beginners often practise on it.

Clusters and Nebulæ. Ursa Major is rich in galaxies, but only a few are reasonably

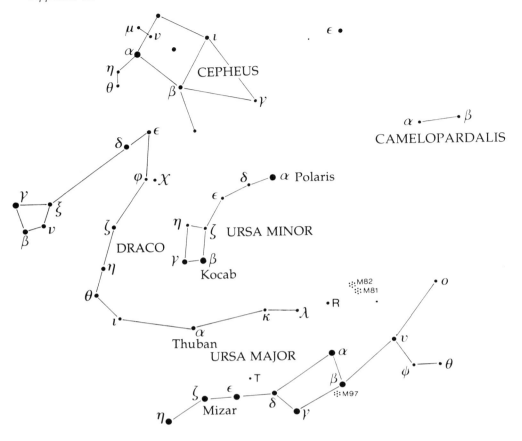

Map III. Ursa Major, Ursa Minor, Draco, Cepheus, Camelopardalis.

bright. Look for M.81 and M.82, which lie side by side; they can be seen with binoculars, close to the star 24 Ursæ Majoris (4.6). M.81 is a spiral, though its form cannot be seen with a small telescope; M.82, a famous radio source, is irregular. The Owl Nebula, M.97, is a planetary; it is not far from Merak, but is difficult to locate with a small telescope because it is decidedly faint. I can see it with a 6-inch reflector, but not at all easily.

Ursa Minor. Chief stars Alpha (Polaris) (2.0), Beta (Kocab) (2.1) and Gamma (3.0). The other stars in the main pattern (Delta, Epsilon, Zeta and Eta) are all well below the fourth magnitude. There are few telescopic objects of note, but the ninth-magnitude companion of Polaris is a good test for a small aperture. Kocab and Gamma are often nicknamed 'the Guardians of the Pole'; Kocab has a K-type spectrum, and is very obviously orange.

Draco. A long, winding constellation, stretching from Lambda (between Dubhe and Polaris) as far as Gamma, which lies near Vega. The chief stars are Gamma (2.2),

Eta (2.7), Beta (2.8), Delta (3.1), Zeta (3.2) and Iota (3.3). Alpha, or Thuban (3.6) used to be the pole star in ancient times.

Doubles. Nu, in the Dragon's 'head', is a wide pair with equal components; keen-sighted people can split it with the naked eye, and it is easy in binoculars. Mu is a much closer pair, again with equal components. Both Epsilon and Eta have faint companions within the range of a 3-inch refractor, but there is nothing particularly notable about them; and it must be said that despite its large area, there is not much of immediate interest in Draco.

Cepheus is not one of the easier constellations to identify, but it is useful to remember that Gamma Cephei lies more or less between Polaris and the W of Cassiopeia. It is better shown in Map VII. The chief stars are Alpha (2.4), Gamma and Beta (each 3.2), Zeta (3.3), Eta (3.4) and Iota (3.5). Telescopic notes are given with Map VII.

Camelopardalis has no star brighter than Beta (4.0). It is a very barren area, lacking in interesting objects, so that it need detain us no further here.

Map IV. Orion, Lepus, Eridanus, Taurus, Cetus, Auriga, Columba, Cælum, Fornax.

The times of rising and setting of Orion, for British latitudes, are given in the notes on Map II. Capella is just circumpolar, but can almost graze the horizon. Perseus is shown in part, and also Triangulum.

Orion.   This may lay claim to being the most glorious of all the constellations. The chief stars are Beta (Rigel) (0.1), Alpha (Betelgeux) (variable, average about 0.5), Gamma (Bellatrix) (1.6), Epsilon (Alnilam) (1.7), Zeta (Alnitak) (1.8), Kappa (Saiph) 2.1), Delta (Mintaka) (2.3, very slightly variable), Iota (2.8), Pi³ (3.2), and Eta and Lambda (each 3.4). Eta is actually a Beta Lyræ-type eclipsing binary with a very small range.

Doubles. Several are given in the list on page 271. Rigel has a 6.8-magnitude companion, easy to see with a 3-inch refractor but (at least to me) very difficult with any smaller aperture; Zeta is easy, Delta even more so. Iota, in the Hunter's Sword, is a fine double, immersed in nebulosity. Theta Orionis is, of course the celebrated Trapezium, and is well worth looking at; Sigma is another multiple, though less striking.

Variables. Betelgeux, naturally, is worth following, but its fluctuations are slow; generally it is brighter than Aldebaran but fainter than Procyon. It is not easy to give a reliable naked-eye estimate, because of the effects of extinction. W Orionis is a very red N-type semi-regular, always within binocular range; U Orionis, easy to locate because it is in the same low-power field as Chi¹ and Chi², is a typical Mira variable with an extreme range of almost eight magnitudes. It reaches maximum only about a week later each year.

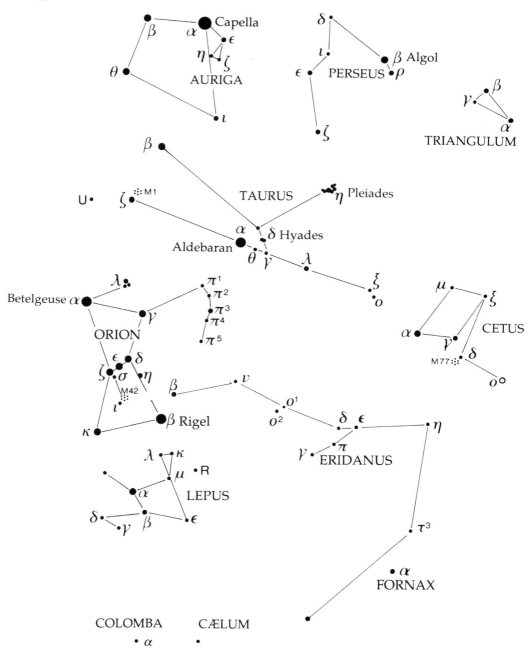

Map IV. Orion, Lepus, Eridanus, Taurus, Cetus, Auriga, Columba, Cælum, Fornax.

Nebulæ. The gaseous nebula M.42 is the finest nebula visible from Britain or the northern States. It is a splendid sight in any small telescope, and dark nebulosity is noticeable near the Trapezium. M.43 is merely an extension of M.42. Close to Alnitak, in the Belt, is the dark nebula nicknamed the Horse's Head because of its undoubted resemblance to the head of a knight in chess. It is not difficult to photograph, provided that the telescope used is of fair aperture and is clock-driven, but visually it is remarkably elusive.

Lepus.   A small but easily-identified constellation south of Orion. The chief stars are Alpha (2.6), Beta (2.8), Epsilon (3.2), Mu (3.3) and Zeta (3.5).

Doubles. For small apertures Gamma is a good, wide pair. Beta and Kappa are less widely separated, but are easy enough; Beta has three more distant companions of magnitudes 10–12.

Variable. R Leporis, near the triangle made up of Mu, Kappa and Lambda, has an N-type spectrum, and is so red that the last-century astronomer Hind nicknamed it 'the Crimson Star'. At its best it reaches naked-eye visibility, and it never becomes very faint.

Cluster. The most interesting cluster in Lepus is a globular, M.79, more or less in line with Alpha and Beta. It is within binocular range, and is in the same field with the star 41 Leporis (5.5). with a 6-inch telescope the outer parts are easily resolved. Its distance from us has been given as 43 000 light-years.

Eridanus.   An immensely long constellation, extending from near Rigel right down to the south polar region; obviously, only part of it can be seen from Britain. The chief stars are Alpha (Achernar) (0.5), Beta (2.8), Theta (Acamar) (2.9), Gamma (3.0) and Delta (3.5). Of these, Achernar and Acamar are invisible from Britian, but are shown in Map XV. Epsilon Eridani, beside Delta, is of magnitude 3.7, and is one of the nearest stars in the sky; it is not too unlike the Sun, though smaller and cooler, and irregularities in its proper motion indicate that is attended by a planet or planets. The part of Eridanus shown on this map has little in the way of interesting objects, though Omicron^2 is a fairly easy double, and the red semi-regular variable Z is within the range of a small telescope even at minimum.

Taurus.   This is one of the most conspicuous of the constellations of the Zodiac. The chief stars are Alpha (Aldebaran) (0.8), Beta (Al Nath) (1.6), Eta (Alcyone) (2.9), Zeta (3.0), Lambda (variable; 3.3 at maximum), Theta^2 (3.4) and Epsilon (3.5). Al Nath was formerly included in Auriga, as Gamma Aurigæ, and its transfer to Taurus seems illogical; it fits in well with the Auriga pattern, while Taurus has no particular shape at all.

Taurus is, of course, dominated by the two open clusters of the Pleiades and the Hyades, but the presence of Aldebaran makes it unmistakable in any case.

Doubles. Both Theta and Sigma, in the Hyades, are so wide that they can be split with the naked eye. Kappa, to the north, forms a very wide pair with 67 Tauri (5.3); the separation is 339 seconds of arc. Aldebaran has a 13th-magnitude optical companion at 121 seconds of arc (position angle 034°), but its faintness makes it a difficult object.

Variables. Lambda, an eclipsing binary of the Algol-type, has a range of little more than half a magnitude, but its fluctuations are easy to follow, and convenient minima are listed in yearly handbooks. T Tauri is the prototype of the class of very young, unstable stars, and is associated with Hind's variable nebula, but the faintness makes it elusive. The usual magnitude of T is about 10.

Clusters and nebulæ. The Hyades are not included in Messier's list, presumably because there was no danger of confusing them with a comet; they are best seen with binoculars, since in a telescope they are too scattered to fit into even a wide-angle field. It is a pity that they are drowned by the brilliant orange light of Aldebaran, which, as we have noted is not a member of the cluster at all, but merely lies in the same line of sight. The Hyades cluster is the closest of all open clusters, and has been of great value to astronomers in distance-measuring. The Pleiades, included by Messier as his No. 45, are more than 400 light-years from us; they too are best seen in binoculars or a very wide-field eyepiece. This is a young group, so that the leaders are hot and blue or bluish-white. The Pleiades contains a beautiful reflection nebula, elusive visually but not hard to photograph.

Then, of course, there is M.1, the Crab, the remnant of the supernova of 1054. I can see it with powerful binoculars (× 12 or over) in the field with Zeta; a telescope shows it clearly, though photography is needed to bring out its complex structure.

Cetus.   Part of this large constellation is shown here; the rest is on Map X. The chief stars are Beta (Diphda) (2.0), Alpha (Menkar) 2.5, and Eta, Gamma and Tau (all 3.5); Tau is one of the closest stars which is at all like the Sun, and has therefore been regarded as a possible candidate for a planetary system. Menkar is obviously orange. The 'head', made up of Menkar, Gamma, Mu (4.3) and Xi (4.4) is fairly distinctive.

Doubles. Gamma is an easy pair for a small telescope. Mira's companion is rather faint, but not difficult when Mira itself is near minimum. I have included the neat little pair 66 Ceti in my list because it acts as a convenient guide to Mira.

Variables. Mira has already been described in the text. It reaches maximum about a month earlier each year, and during the late 1980s these maxima occur at a convenient time, so that the star is worth watching.

Galaxies. Cetus is fairly rich in galaxies. Much the most interesting is M.77, a giant spiral near Delta which is a powerful radio source. Its magnitude is only just above 9, but it is worth finding even though small telescopes show it as nothing more than a dim patch. M.77 has the distinction of being the largest and most luminous galaxy in Messier's list.

Auriga.   An imposing constellation of distinctive shape. The chief stars are Alpha (Capella) (0.1), Beta (Menkarlina) (1.9), Theta (2.6), Iota (2.7), Epsilon (variable; 3.0 at maximum) and Eta (3.2). Al Nath also forms part of the pattern, but, as we have noted, has been purloined by Taurus.

Capella is shown in both key maps, and is surpassed by only three other stars visible from Britain: Sirius, Arcturus and Vega. The difference between Vega and Capella is only 0.05 of a magnitude, so that whichever of the two is the higher will also appear the brighter.

Doubles. Theta is said to be a test for a 4-inch refractor, but I always find it rather difficult with any aperture below 6 inches. Omega is a neat pair, and a useful test.

Variables. Both Epsilon and Zeta, in the triangle of the Hædi or 'Kids', are giant eclipsing binaries, described in the text; but Zeta's fluctuations are slight, and Epsilon will not again show an eclipse for many years. The Mira-type variable R Aurigæ can exceed the seventh magnitude; it has a long period of over 457 days, but this period is notoriously inconstant.

Clusters. Auriga is crossed by the Milky Way, and there are rich star-fields here; look too for the three bright open clusters M.36, 37 and 38, all of which are in the same field with × 7 binoculars. Even a small telescope will resolve them well.

Columba.   An unremarkable constellation south of Lepus; it is always very low from England, and part of it does not rise at all. The chief stars are Alpha (2.6) and Beta (3.1). Columba contains nothing of immediate interest.

Cælum.   A very obscure and dull constellation, always very low from Britain; there is no star above magnitude 4.4. Gamma (4.5) is a reasonably easy double.

Fornax.   Also very low and obscure; its brightest star, Alpha, is only of magnitude 3.9. Fornax is rich in galaxies, but all these are too faint to be properly seen with small apertures.

Map V. Gemini, Cancer, Canis Major, Canis Minor, Monoceros, Hydra, Puppis.

The constellations shown in this map are at their best during winter and spring evenings. The following times of rising and setting for England are for Cancer, and are of course very rough. Gemini, Cancer and Leo are all included in the Zodiac.

1 January:	Rises 6 p.m., highest 2 a.m., sets in daylight.
1 April:	Rises in daylight, highest 8 p.m., sets 4 a.m.
1 July:	Rises in daylight, highest in daylight, sets in daylight.
1 October:	Rises at midnight, highest 8 p.m., sets in daylight.

Canis Major is most notable because of the presence of Sirius, much the brightest star in the sky. The other chief stars are Epsilon (1.5), Delta (1.9), Beta (2.0, very slightly variable), Eta (2.4) and Zeta and Omicron² (each 3.0). It is interesting to note that Delta, Eta and Omicron² are all extremely luminous, and Delta may be at least 130000-times as powerful as the Sun, while the nearby Sirius, at 8.6 light-years, is a mere 26 Sun-power!

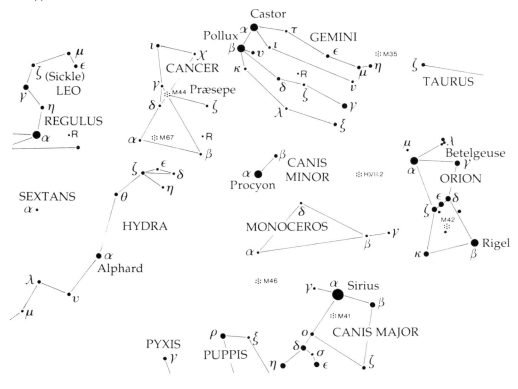

Map V. Gemini, Cancer, Canis Major, Canis Minor, Monoceros, Hydra.

Doubles. The Companion of Sirius is not particularly faint, but the overpowering brilliance of the primary makes it a difficult object. Epsilon is much easier, and a small telescope will show both stars well.

Variable. There are no bright Mira stars in Canis Major, but it is worth looking at R, which is an Algol-type eclipsing binary with a range of just over half a magnitude and the short period of only 1.14 days.

Cluster. There is one Messier object in the constellation. This is the rich open cluster M.41, close to the reddish star Nu² (3.9) not far from Sirius. It is easily detectable with the naked eye, and a low power on a small telescope will show it well. There are, of course, some good star-fields in Canis Major, which is crossed by the Milky Way.

Canis Minor. The Hunter's junior Dog contains Alpha or Procyon (0.4); the other bright star is Beta (2.9). Beta makes a neat little triangle with Gamma (4.3) and Epsilon (5.1). Nearby is S Canis Minoris, the brightest of several Mira-type variables in the constellation, which can rise to magnitude 6.6 at maximum. There is little else of interest here.

Monoceros. A large, faint constellation within the triangle formed by Sirius,

Procyon and Betelgeuse. The leading star (Beta) is only of magnitude 3.7, but there are some good, rich fields.

Doubles. Beta is a good example of a multiple star, with a main pair (AB) and two more distant companions; it is worth looking at. Epsilon is a fairly wide, easy pair.

Variables. U Monocerotis is notable as being a member of the comparatively rare RV Tauri class, with alternate deep and shallow minima; the full period is 92.3 days, and the star never fades much below the eighth magnitude. The constellation also includes several Mira variables, one of which, V, can reach magnitude 6.

Clusters and nebulæ. Monoceros is crossed by the Milky Way, and is rich in clusters – including M.50, which is visible in binoculars and easily resolvable with a 3-inch refractor. The cluster NGC 2244, round the star 12 Monocerotis (5.8) is also a binocular object, but its main interest lies in the fact that it is surrounded by the Rosette Nebula (NGC 2237–9), which is superb photographically but not easy to see visually. Two other nebulæ – NGC 2261, round R Monocerotis, and NGC 2264, the Cone Nebula round the multiple S Monocerotis – require fairly large apertures. NGC 2261, also known as Hubble's Variable Nebula, shows marked alterations in brightness, and I admit that I have never yet been able to see it with my 15-inch reflector.

Puppis. This is the Poop of the now-dismembered constellation of Argo Navis. It is shown in Map XIV, but a few of its stars are sufficiently far north to rise from Britain. Look for the lovely open cluster M.46, which forms a triangle with Rho Puppis (2.8) and Sirius.

Gemini. A magnificent constellation, crossed by the Milky Way and abounding in rich fields and interesting objects. The leading stars are Beta (Pollux) (1.1), Alpha (Castor) (1.6), Gamma (Alhena) (1.9), Mu (2.9), Epsilon (3.0), Eta (variable; 3.1 at maximum), Xi (3.4) and Delta (3.5). Castor and Pollux can be found from either key map. Pollux, a K-type orange star, is appreciably brighter than Castor.

Doubles. Castor is, of course a fine binary, and it is not a difficult object with a small telescope. The faint third member of the group is an eclipsing binary known officially as YY Geminorum. Delta is usually regarded as a test for a 2-inch refractor, but I find it easy with such an aperture.

Variables. The two naked-eye variables, Eta and Zeta, have already been described. Eta is semi-regular, but though the official period is around 233 days I have never been convinced! Zeta is a typical Cepheid. Of other variables, R is a Mira star which can attain the sixth magnitude. U Geminorum is much fainter, and is the prototype dwarf nova, though many observers refers to the class as of the SS Cygni type. U Geminorum flares about every 103 days, but it is always faint, and you will need a set of specialist charts to locate it.

Clusters and nebulæ. Gemini contains one splendid open cluster, M.35, easy to find because it is close to Eta and the star 1 Geminorum (4.2). Any telescope will resolve it into stars. Owners of larger apertures may like to seek out the Eskimo Nebula, a planetary, but it is only of the tenth magnitude, and is decidely elusive. Its position is 07^h 29 m·2, $+ 20°55'$.

Cancer.   A Zodiacal constellation, but very dim; it looks not unlike a faint and ghostly version of Orion. Its brightest star is Beta (3.5).

Doubles. Zeta is a splendid example of a multiple star; the two main components are easy to separate, but the closer pair is much more difficult. There is a faint, more distant companion. With the naked eye, Zeta looks like a single star of magnitude 4.7.

Variables. R Cancri is a normal Mira star with a period of four days less than a year. The semi-regular X Cancri is intensely red, and is worth finding; as it never fades below magnitude 7.5, it remains within range of binoculars.

Clusters. Cancer is redeemed by the presence of two very important open clusters. M.44, Præsepe, is inferior only to the Pleiades and Hyades so far as British observers are concerned; it is easy to locate with the naked eye, flanked to either side by the two Aselli or 'Asses', Gamma Cancri (4.7) and Delta (3.9). Præsepe is easily resolvable, and with a wide-field eyepiece it is a magnificent sight. M.67, near the star Acubens or Alpha Cancri (4.2) is on the firnge of naked-eye visibility; the French astronomer Camille Flammarion described it as resembling 'a sheaf of corn'. M.67 is famous because it seems to be one of the oldest known open clusters; it has persisted because it lies well away from the main plane of the Galaxy, and has not been torn asunder by passing non-cluster stars.

Hydra.   With the demise of Argo Navis, Hydra is now the largest of the 88 recognized constellations; parts of it are also shown on Maps VI and IX. It is, however, rather barren. The chief stars are Alpha (Alphard) (2.0), Gamma (3.0), Zeta and Nu (each 3.1), Pi (3.3), Epsilon (3.4) and Xi (3.5). Alphard is shown on Key Map II; it is easy to find, as it is distinctly reddish, and its nickname of 'the Solitary One' suits it well. It can be identified by continuing the sweep from Bellatrix through Betelgeuse and Procyon; also, Castor and Pollux point to it. It is a fine sight in a low power.

Doubles. Theta is a useful test, because the companion is rather faint. Epsilon, in the Watersnake's Head, is triple; there are two main components which are not very unequal.

Variables. Hydra contains two red variables. R is a Mira star (shown on Map VI) which can rise to the fourth magnitude; its present period is around 390 days, but in the eighteenth century was as much as 500 days, so that it may still be shortening. U Hydræ, a semi-regular with a range of about one magnitude, has an N-type spectrum, and is very red.

Sextans.   A very dim constellation, with no star brighter than Alpha (4.5). There is nothing of note here.

Parts of Leo and Taurus are included in this map, but are better shown in Maps VI and IV, respectively.

Map VI. Leo, Virgo, Coma Berenices, Corvus, Crater, Leo Minor.

This area contains some interesting objects; Regulus, Spica and Arcturus are shown in Key Map I. The rough times of rising and setting for Spica, in England, are:

1 January: Rises 1 a.m., highest 6 a.m., sets in daylight.
1 April: Rises 7 p.m., highest midnight, sets 5 a.m.
1 July: Rises in daylight, highest in daylight, sets 11 p.m.
1 October: Rises in daylight, highest in daylight, sets in daylight.

Leo.   A large and important Zodiacal constellation. Regulus (Alpha), of magnitude 1.3, is the leader, and is in fact the faintest star to be officially ranked as of the 'first magnitude'. Next come Gamma (2.0), Beta (2.1), Delta (2.6), Epsilon (3.0), Theta (3.3), Zeta (3.4) and Eta and Omicron (each 3.5). The curved line of stars beginning at Regulus is known as the Sickle, and is very conspicuous; the triangle formed by Beta, Delta and Theta is also unmistakable. Beta (Denebola) was ranked as of the first magnitude in ancient times, but is now below the second. As it is suspected of current variability, it is worth watching. Gamma makes a good comparison star.

Doubles. Gamma Leonis (Algieba) is a magnificent binary – one of the finest stellar objects in the sky for a small telescope, particualrly as the primary is an orange-red star of type K. It is a binary, but the period is over 600 years, so that there is little relative motion. There are two faint, more distant companions, unrelated to the main pair. Omega is a neat double, and Regulus has a faint optical companion at a distance of 177 seconds of arc.

Variable. R Leonis, near the pair of stars 18 and 19 Leonis, is a Mira variable with a period of 312 days. Since it never drops much below magnitude 11, and at maximum can reach naked-eye visibility, it is easy to identify, particularly as – like all Mira stars – it is red. It makes up a conspicuous small triangle with 18 and 19.

Galaxies. Faint galaxies abound in Leo, and there are five Messier objects: 95, 96, 105, 65 and 66. Of these the brightest are M.65 and 66, near Theta, which can be seen with a small telescope. Like the rest, they appear as blurred patches, but are of obvious interest to supernova-hunters.

Leo Minor.   A small constellation adjoining Leo and Ursa Major. It has no star above magnitude 3.8. Beta is a close double which is a good test; the Mira variable R Leonis Minoris can reach magnitude 6.3 at its best.

Virgo.   In shape Virgo is rather like a roughly-drawn Y. The leading stars are Alpha (Spica) (0.9), Gamma (Arich) (2.7), Epsilon (2.8), and Zeta and Delta (each 3.4).

Doubles. Gamma (Arich) is a splendid binary, with a period of 171 years. It used to be very wide and easy; it is still not difficult, but is closing, and by the year 2110 will be a very difficult object. The components are exactly equal. Theta Virginis has a

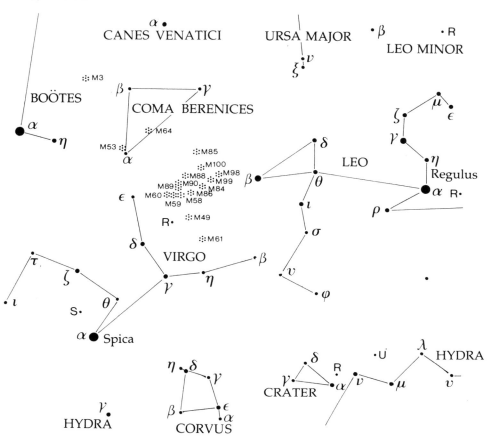

Map VI. Leo, Virgo, Coma Berenices, Canes Venatici, Corvus, Crater, Leo Minor.

companion of magnitude 9.4, at a distance of 7 seconds of arc, making it a useful test object for small apertures.

Variables. Virgo is not rich in variable stars, but R and S, both Mira stars, are easy objects with a small aperture when near maximum.

Galaxies. The interior of the 'bowl' of Virgo – enclosed by the line from Epsilon to Beta, and on the far side by Beta Leonis – is crowded with galaxies, and is worth sweeping. The Messier objects are 61, 84, 86, 49, 87, 88, 89, 90, 58, 59, 104 (the 'Sombrero Hat'), 59 and 60. Of these, the most interesting is M.87, also known as Virgo A because it is an exceptionally powerful source of radio waves. It is a giant elliptical system, from which issues a curious jet of material visible only with large telescopes. It has a bright centre, and since the integrated magnitude is above 9 it is not hard to locate. Also of special note is M.104, which is a fine sight in a moderate telescope; with a 12-inch reflector it is possible to see the dark absorption band running across the galaxy which has led to its being nicknamed the Sombrero Hat.

Coma Berenices. This whole area gives the impression of a large, extended star-cluster; there are no bright stars, the leaders (Alpha, Beta and Gamma) all being of magnitude 4.3. Coma is swarming with galaxies; Messier objects are numbers 64, 85, 88, 98, 99 and 100. M.64, the 'Black-Eye' Galaxy, is located close to the fifth-magnitude star 35 Comæ, and is an easy object, though the dark 'black-eye' area north of the centre cannot be seen with a telescope below 10 inches in aperture, at least to an observer no more keen-sighted than I am. Note also the globular cluster M.53, a degree north-east of Alpha Comæ, which in my view is one of the most beautiful of all globulars. A 6-inch reflector shows it excellently.

Canes Venatici. The little constellation of the Hunting Dogs adjoins Ursa Major, Boötes and Coma. Its only bright star is Alpha (2.9), known as Cor Caroli, which is a wide, easy double; any telescope should split it.

Variables. There are several Mira stars in Canes Venatici, of which the brightest, R, can rise above magnitude 7. The semi-regular Y Canum Venaticorum has an N-type spectrum, and is so red that it has been named La Superba. It is usually within binocular range, and can always be followed with a 3-inch refractor.

Clusters and nebulæ. M.3 is a lovely globular, and is easy to locate; binoculars will show it, and a 6-inch telescope will resolve all but its inner parts. It lies near the boundary of Canes Venatici, and the best guide to it is the star Beta Comæ. The whole constellation teems with galaxies, including M.106, 94, 63 and 51. M.51, the Whirlpool Galaxy, was actually the first to have its spiral form recognized (by Lord Rosse, in 1845), and with my 15-inch reflector I can see the spiral unmistakably. It is linked with its neighbour galaxy, NGC 5195, by a sort of 'arm' which can be traced with apertures as small as 12 inches. M.51 is about $3\frac{1}{2}$ degrees from Alkaid, in the Great Bear, which is the best guide to it. I am assured that it can be glimpsed with binoculars, but I have never been able to see it without a telescope.

Boötes. This constellation is shown in part, but is described with Map IX.

Hydra. Hydra is also partly shown, the brightest star being Nu (3.1). Note the N-type variable U Hydræ, which is intensely red.

Corvus. This little constellation is surprisingly conspicuous, because its four leaders make up a well-marked quadrilateral; they are Gamma and Beta (each 2.6) and Delta and Epsilon (each 3.0). Alkhiba, which is lettered Alpha Corvi, is a full magnitude fainter. There are few objects of note, but R is a Mira-type variable which can attain magnitude 6.7, and Delta is a fairly easy double.

Crater adjoins Hydra, close to Nu Hydræ; its brightest star, Delta, is only of magnitude 3.6. It contains nothing of real interest to the owner of a small or moderate-sized telescope.

Map VII. Cassiopeia, Cepheus, Lacerta, Perseus, Andromeda, Lynx, Triangulum.

Of the groups in this map, all are circumpolar from Britain apart from sections of Andromeda and Triangulum. Ursa Minor and Camelopardalis are also shown, but are described with Map III.

Cassiopeia, shown in Key Map I, is one of the most interesting and important of the far-northern constellations. The Milky Way passes through it, and there are many rich telescopic fields. The chief stars are Gamma (variable, usually about 2.2), Alpha or Shedir (2.2, but suspected of slight variability), Beta (2.3), Delta (2.7), Epsilon and Eta (each 3.4) and Zeta (3.7). The first five make up the familiar pattern.

Doubles. Eta and Iota are both easy binaries of long period; Lambda has two very nearly equal components, separated by half a second of arc. Alpha (Shedir) has a ninth-magnitude companion at a distance of 63 seconds of arc, but this is an optical pair; the two components are not genuinely associated.

Variables. Gamma is an unpredictable star, and is unstable. At its brightest (as in 1936) it can reach magnitude 1.6; by 1939 it was down to 3.2; for many years now it has fluctuated around 2.2, but it may decide to 'perform' again at any time. Beta makes a good comparison star. Shedir or Alpha Cassiopeiæ is suspected of slight variability, and my observations of it, which go back to 1935, indicate that it has a range of between 2.0 and 2.5, though again the usual magnitude is about 2.2. It has a K-type spectrum, and is obviously orange. The unique Rho Cassiopeiæ has been described in the text, and is worth watching. Of telescopic variables in the constellation, the brightest is R, a typical Mira star with a long period. Some maxima may reach magnitude 4.7, but at minimum the magnitude is well below 13, beyond the reach of small telescopes.

Clusters and nebulæ. Cassiopeia is immersed in the Milky Way, and as well as rich star-fields there are various open clusters including two Messier objects, M.52 and M.103. M.52 is prominent and easy to resolve, but M.103 is poor and sparse, and it is hard to see why Messier listed it and omitted the more prominent cluster NGC 663. NGC 457 is of special interest because the faint Phi Cassiopeiæ lies at its south-east edge. Whether or not Phi is a true cluster member is not certain; if it is, it must be at least 200 000 times as luminous as the Sun since the cluster is over 9000 light-years away.

Cepheus. This is not too easy to identify. The chief stars are Alpha (2.4), Gamma and Beta (3.2), Zeta (3.3), Eta (3.4) and Gamma (3.5). Gamma lies betwen Beta Cassiopeiæ and Polaris; the main part of the constellation is located between Cassiopeia and Vega. The triangle made up of Zeta, Delta and Epsilon is the most easily-recognizable feature.

Doubles. Beta, Kappa and Xi are all easy; the famous variable Delta has a 7.5-magnitude companion at a distance of 41 seconds of arc.

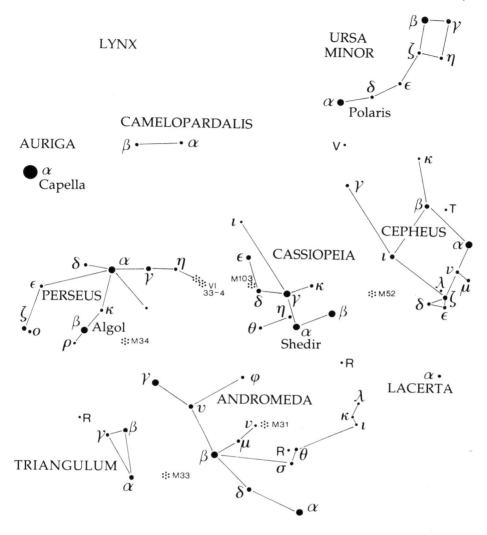

Map VII. Cassiopeia, Cepheus, Lacerta, Perseus, Andromeda, Lynx, Triangulum.

Variables. Delta Cephei is the prototype of the Cepheid class, and with its range of from 3.5 to 4.4. it is always a naked-eye object; Zeta and Epsilon make good comparisons. Mu Cephei, Sir William Herschel's 'Garnet Star', can rise to 3.4, but may also fall below 5. It has an M-type spectrum, and, as its nickname shows, it is extremely red. It is a highly luminous supergiant, and is actually much more powerful than Betelgeux, though it looks fainter because it is much further away. The colour can be distinguished with the naked eye, but in a telescope it is pronounced. Good comparison stars are Nu (4.4) and Epsilon. T Cephei is a typical Mira star; S Cephei,

301

with an N-type spectrum, has a long period (487 days) and is one of the reddest stars in the sky. To me it looks just as impressive as Hind's 'Crimson Star', R Leporis.

Lacerta adjoins Cepheus, but has only one star (Alpha, 3.8) above the fourth magnitude. The only object of any immediate interest is the open cluster NGC 7243, which forms a triangle with Beta (4.4) and Alpha. It is not hard to resolve, but it is in no way remarkable.

Perseus.   Here we have a very prominent group. The chief stars are Alpha or Mirphak (1.8), Beta or Algol (2.1 at maximum), Zeta (2.8), Epsilon (2.9), Gamma (2.9), Delta (3.0), and Rho (about 3 at maximum). Zeta, Epsilon and Gamma all have faint companions.

Variables. Algol is the prototype eclipsing binary, and is fascinating to watch during its minima. Rho Persei is a red semi-regular with a range of from about 3 to 4; I have never had much faith in the 'official' period of about 50 days. The best comparison is Kappa (3.8), on the opposite side of Algol. X Persei, near Zeta, seems to be irregular, and has a spectrum of type 09.5; it is an X-ray source, and will repay attention, particularly from binocular-users.

Clusters and nebulæ. The main clusters in Perseus are those making up the Sword-Handle (not to be confused with the Sword of Orion), NGC 869 and 884 — Herschel's H.VI, 33–4. They are in the same low-power field, and in my view are among the most beautiful objects in the sky; between them is a faint red star. The two can be seen with the naked eye as a dim blur, but any telescope will resolve them fully. Other open clusters are M.34, near Algol, and NGC 1528, near Mirphak. Much more elusive is the gaseous California Nebula, NGC 1499, which has low surface brightness but is well worth finding.

Andromeda leads off the Square of Pegasus in the direction of Perseus; Alpheratz, now known as Alpha Andromedæ, used (more logically) to be referred to as Delta Pegasi. The three leaders of Andromeda — Beta, Alpha and Gamma — are all of magnitude 2.1; next comes Delta (3.3). Beta, or Mirach, is of type M, and is obviously orange; so too is Gamma or Almaak, of type K.

Doubles. Gamma is a fine, easy double, with an orange primary and a companion which looks bluish by contrast. The companion is itself a double, with rather unequal components and a separation of 0.5 seconds of arc. I can split it with my $12\frac{1}{2}$-inch reflector, but I always find it difficult with smaller apertures; it is a good test object.

Variables. There are several fairly bright Mira stars in Andromeda; pride of place must go to R Andromedæ, which has a long period and a very considerable range (5.8 to 14.9). Z Andromedæ is the prototype of its class, but never rises above magnitude 8.

Clusters and nebulæ. M.31, the Great Spiral, is the nearest of the large galaxies, at a distance of 2.2. million light-years; it is visible with the naked eye, and binoculars show it well, but not even a large telescope will make it look impressive, because it lies at a narrow angle to us. In the same low-power field are its two companions, M.32 and NGC 205. The nearest naked-eye star is Nu (4.5).

Triangulum. For once we have a constellation which resembles the object after which it is named; the three brightest stars – Beta (3.0), Alpha (3.4) and Gamma (4.0) do indeed make up a triangle. It is easily found, near Beta and Gamma Andromedæ.

Variable. R Trianguli is a Mira star, easy to locate when at maximum but inconveniently faint at minimum, when it drops almost to the thirteenth magnitude.

Galaxy. M.33, a member of the Local Group, is a rather loose spiral, roughly between Alpha Trianguli and Beta Andromedæ. It has been claimed that it can be glimpsed with the naked eye, and it is very easy in binoculars, but it is surprisingly difficult to locate with a telescope, because its surface brightness is so low. I recommend finding it with binoculars and then turning to your telescope, though I am afraid you will see it as little more than a dim blur.

Lynx. One of the most barren of all constellations – it is said to have been named because lynx-like eyes are needed to show anything there! There is, however, one fairly obvious star, Alpha (3.1). The Mira variable R can reach almost to the seventh magnitude, but at minimum fades to below 14, and without setting circles it is hard to identify, because it lies in so barren an area. Lynx contains nothing else of note.

Map VIII. Cygnus, Lyra, Sagitta, Vulpecula, Delphinus, Equuleus, Capricornus, Aquila, Sagittarius, Scutum, Serpens, Aquarius.

This area is dominated by Vega, Altair and Deneb, which make up a large triangle; years ago I nicknamed it 'the Summer Triangle', and everyone now seems to use the term, though it is unofficial, and in any case does not apply to the southern hemisphere, where it should more properly be called 'the Winter Triangle'. Vega and Deneb are just circumpolar from England; the times of rising and setting for Altair are given below. Remember that in all these rising and setting tables, allowance must be made for British Summer Time.

1 January:	Rises 6 a.m., highest in daylight, sets 8 p.m.
1 April:	Rises at midnight, highest 7 a.m., sets in daylight
1 July:	Rises in daylight, highest 1 a.m., sets in daylight.
1 October:	Rises in daylight, highest 7 p.m., sets 2 a.m.

Vega and Deneb are shown on Key Map I. Vega is almost overhead at midnight near midsummer, and can be recognized by its brilliance and its bluish colour, differing markedly from the yellowish hue of Capella, which occupies the overhead position at times during the winter.

Lyra. Though Lyra is a small constellation, and Vega is the only star above the third magnitude, it has more than its fair share of interesting objects. The leading stars are

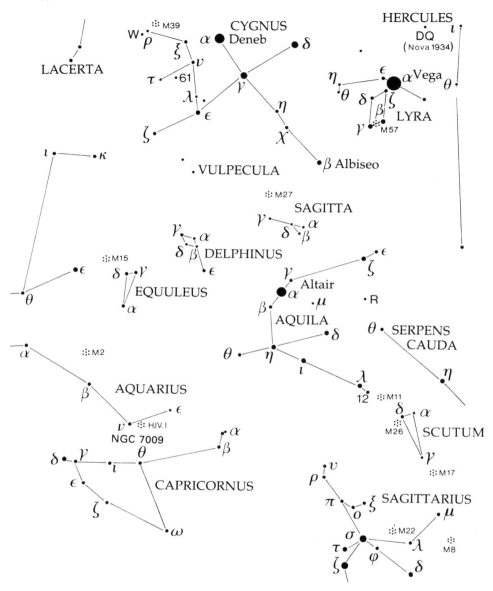

Map VIII. Cygnus, Lyra, Sagitta, Vulpecula, Delphinus, Equuleus, Capricornus, Aquila, Sagittarius, Scutum, Serpens, Aquarius.

Alpha or Vega (0.0), Gamma (3.2) and Beta (variable, 3.4 at maximum). The quadrilaterial made up of Beta, Gamma, Delta and Zeta is very distinctive, even though Delta and Zeta are both below the fourth magnitude.

Doubles. Epsilon, the famous double–double or quadruple star, has been described in the text. The two main components can be split with the naked eye, and a 3-inch refractor will show both as doubles; Epsilon¹ (AB) is slightly easier than Epsilon² (CD). Zeta is also a wide, easy double, very obvious with a small telescope. Delta is yet another double, with a separation of over 10 minutes of arc; the brighter component is a red giant of type M, while the fainter member of the pair is white. Binoculars will separate them, and in a wide-field telescope they make a lovely spectacle, particularly as the whole field is very rich in fainter stars.

Variables. Beta Lyræ is, of course, the prototype of its class of eclipsing binaries, and is easy to follow, particularly because its neighbour Gamma makes an excellent comparison star. R Lyræ is a semi-regular, best studied with binoculars because in this case there is a singular lack of convenient comparisons. There is also the prototype of the RR Lyræ stars, but it never rises as high as magnitude 7, and is not easy to locate without a set of detailed charts.

Clusters and nebulæ There are two interesting nebular objects in Lyra. One is the globular cluster M.56, which is on the border of the constellation; the nearest bright naked-eye star is Beta Cygni. M.56 is easy to see with a small telescope, but it does not have the marked central condensation so often found with globulars. M.57, the Ring Nebula, is a planetary, lying midway between Beta and Gamma. A 3-inch refractor will show it as a tiny, luminous ring, but I doubt whether any aperture below 12 inches will bring out the faint central star, and I admit that I have never seen it clearly with anything below 15 inches.

Cygnus. The Swan, but also, and more appropriately, known as the Northern Cross. It is a superb constellation, in a particularly rich part of the Milky Way. The chief stars are Alpha (Deneb) (1.2), Gamma (2.2), Epsilon (2.5), Delta (2.9), Beta (3.1) and Zeta (3.2). It is worth noting that Beta, often still known by its proper name of Albireo, lies roughly between Vega and Altair.

Doubles. Albireo is, I maintain, the most beautiful double in the sky. It is so wide and easy that any telescope will split it – so, for that matter, will really powerful binoculars – and the colours are striking, with a golden-yellow primary and a companion which some people describe as blue and others as green (personally, I opt for blue). Delta is also an easy double, and there is also 61 Cygni, which has the distinction of being the first star to have its distance measured – which makes it worth finding, even though it is dim and unspectacular.

Variables. Cygnus abounds with variables. Chi Cygni, close to Eta, has an S-type spectrum, so that it is intensely red; it has the greatest range of any Mira star, and has been known to approach the third magnitude, though at most maxima barely equals its neighbour Eta (3.9). At minimum it is not easy to identify, because it lies in so rich an area. My advice is to find it when it is bright, and then check the field, memorizing it so that you can follow it through its cycle – but you will need a fair-sized telescope, because at minimum Chi drops below the fourteenth magnitude. Other very red Mira stars are U Cygni and R Cygni; W, a semi-regular, is a good binocular object, and so is X, a typical Cepheid. SS Cygni, the prototype dwarf nova, is not hard to locate if you have a set of charts, and when at its brightest it is within the range of a 3-inch refractor. P Cygni, near Gamma, was once ranked as a nova; it is an unstable star, but for many years now it has hovered around the fifth magnitude.

Clusters and nebulæ. Here, too, Cygnus is rich, and this is also the region of dark rifts in the Milky Way, which are striking with binoculars. Look too for the North America Nebula (NGC 7000), near Deneb, and two open clusters, M.29 and M.39. M.39, near Rho Cygni, is a fine sight in a small telescope.

Vulpecula.   The brightest star (Alpha) is only of magnitude 4.4; it makes a very wide pair with 8 Vulpeculæ, of magnitude 5.8, and since the separation is well over 400 seconds of arc the best view is obtained with binoculars or a very wide-field eye-piece. The main object of interest in Vulpecula is the Dumbbell Nebula, one of the finest of all planetaries, which lies close to Gamma Sagittæ in the neighbouring constellation of the Arrow. It is Messier's 27th object, and was discovered by him as long ago as 1764. Its nickname is obvious from its shape, which can be made out unmistakably with a 6-inch reflector. The central star is not easy, but I have seen it with my 12½-inch reflector.

Sagitta.   A compact little group, quite unmistakable even though its stars are not bright; the leaders are Gamma (3.5) and Delta (3.8). Gamma, as we have noted, is the best guide to M.27, the Dumbbell Nebula in Vulpecula.

Double. Zeta Sagittæ is an easy pair. The brighter component is a binary with a period of just under 23 years; it is fairly close (0.3 seconds of arc) and is a good test object.

Variables. There are two fairly bright short-period variables, S (a Cepheid) and U (an Algol-eclipsing binary), but the main interest is centred on WZ, which is a recurrent nova. Its usual magnitude is below 15, but on three occasions – in 1913, 1946 and 1978 – it has risen briefly to magnitude 7, and it is always worth watching. It may possibly show another outburst around the year 2011!

Equuleus.   A small constellation; the brightest star is Alpha (3.9). The main object of interest is Delta, which is a rapid binary; its period is under 6 years. The components are almost equal, but the separation is never more than 0.3 seconds of arc, so that it is a difficult object.

Pegasus.   Most of this constellation, including the Square, is shown on Map X. The chief star in the present map is Epsilon or Enif (2.4), which has been suspected of variability; it is of type K, and distinctly orange. Near it is the fine globular cluster M.15, which is not greatly inferior to M.13; it has a condensed centre, but the outer parts are easy to resolve.

Aquarius. The Water-bearer also lies mainly in Map X. Two of its main stars, Beta and Alpha, are on the present map, together with two nebular objects; the fine globular M.2, which I can resolve fully with a 12½-inch reflector and which lies between Beta Aquarii and Gamma Pegasi, and the beautiful planetary NGC 7009, which is in the same low-power field with the orange star Nu Aquarii (4.5). NGC 7009 is often nicknamed the Saturn Nebula; its distinctive shape can be made out with a 10-inch telescope, but smaller apertures will show it easily as a patch of light.

Aquila. Another splendid constellation, crossed by the Milky Way. The chief stars are Alpha (Altair) (0.8), Gamma (2.7), Zeta (3.0), Theta (3.2) and Delta and Lambda (each 3.4). Altair is flanked to either side by Beta (3.7) and Gamma; Gamma is of type K, and its orange hue is unmistakable. The line made up of Theta, Eta and Delta is easy to identify; Eta is the celebrated Cepheid variable. Its fluctuations were discovered in 1784, only a few months after those of Delta Cephei. Had they been noticed slightly earlier, the short-period variables would probably have become known as Aquilids rather than as Cepheids. The other fairly bright variable in Aquila is R, a Mira star which can attain magnitude 5.5 and never drops below 12. There are several fairly prominent open clusters, but, surprisingly, no notable doubles.

Serpens. This strange constellation is divided into two parts, Caput (the Head) and Cauda (the Body), separated by Ophiuchus. Caput is shown in Map IX. The brightest stars in Cauda are Eta (3.3) and Theta or Alya (3.4). Alya is one of my favourite doubles; it is very wide and easy, and I can just about split it with × 20 binoculars. To find it, simply follow along the line made up of Theta, Eta and Delta Aquilæ.

Also in Cauda is M.16, the Eagle Nebula, which is made up of a bright cluster together with gaseous nebula. It is only just inside Serpens, and the best guide to it is Gamma Scuti (4.7). The cluster is bright, but the nebulosity is much more elusive, and I have never seen it easily with any aperture much below 12 inches. Photographs are needed to show it really well.

Scutum. Scutum, lying beyond the 'end' stars of Aquila (Lambda and 12) has no star brighter than Alpha (3.8), but it lies in a very rich part of the Milky Way. There are two objects of special note: R Scuti, the brightest of all the RV Tauri variables, and the lovely cluster M.11. R Scuti is an excellent binocular object, particularly as there are suitable comparison stars nearby, and it seldom falls below the range of binoculars. M.11 is often known as the Wild Duck – a name due to the last-century astronomer Admiral W.H. Smyth, who likened it to 'a flight of wild ducks', but to me it gives an impression of a fan. It is a superb sight in a small telescope, and is very easy to locate; it is visible with the naked eye, though care must be taken in picking it out from the surrounding rich area.

Sagittarius. This is a large and bright Zodiacal constellation, but unfortunately it is always very low over Britain, and the southernmost part of it never rises; all of it is shown in Map XIII. The chief stars are Epsilon (1.8), Sigma (2.0), Zeta (2.6), Delta (2.7), Lambda (2.8), Pi (2.9), Gamma (3.0), Eta (3.1), Phi (3.2), Tau (3.3) and Xi² (3.5). Curiously, the stars lettered Alpha and Beta are comparatively faint, at magnitudes 4.0 and 3.9 respectively.

The Milky Way is particularly rich here; the glorious star-clouds indicate the direction of the centre of the Galaxy, and the whole region will repay sweeping with binoculars or a low magnification on a telescope. British-based observers cannot really appreciate how magnificent Sagittarius really is.

Doubles. In the section shown in the present map, the most impressive double is Zeta, which has almost equal components and a period of 21 years. It is fairly close (0.3 seconds of arc) but is not really a difficult object.

Variables. There are several fairly bright Mira stars, notably R and RR, but the most interesting variable is RY Sagittarii, which is of the R Coronæ Borealis type and is the brightest member of the class apart from R Coronæ itself. Its usual magnitude is 6, but at minimum it falls to 15. Unfortunately its southerly declination makes it hard to follow by British observers.

Clusters and nebulæ. Sagittarius contains more globular clusters than any other constellation, including M.75, 28, 69, 22, 54, 70 and 55, all of which will repay attention. Open clusters in Messier's list are M.23, 18, 20 and 25; note that M.24 is not a true cluster, but merely a star-cloud in the Milky Way. Gaseous nebulæ include M.17 (the Omega Nebula), M.20 (the Trifid) and M.8 (the Lagoon), which are among the finest of their type, even though photography is needed to show their full beauty. G.F. Chambers descibed M.17 as looking like 'a swan floating on water', and some people still call it the Swan Nebula. Open clusters include three Messier objects: 23, 18 and 25, all of which are easily resolvable with small telescopes.

Capricornus. Capricornus is a Zodiacal constellation, but rather barren. The brightest stars are Delta (2.9) and Beta (3.1). There is not much of immediate interest in Capricornus, but Alpha is a wide naked-eye double – though the components are not genuinely associated; the fainter member is 1600 light-years away, while the brighter lies at only 116 light-years. There is one nebular object, the globular cluster M.30, not far from Zeta (3.7). The outer parts are easy to resolve, but the centre is condensed. It is in the same low-power field with the star 51 Capricorni.

Hercules. A small part of Hercules appears in Map VIII, but most of the constellation lies in Map IX, and is described there.

Map IX. Boötes, Corona Borealis, Hercules, Serpens, Ophiuchus, Libra, Scorpius.

These are mainly summer groups, though the northernmost parts of Hercules and Boötes are circumpolar in England. Rough times of rising and setting for Antares, in Scorpius, are:

1 January: Rises 5 a.m., highest in daylight, sets in daylight.
1 April: Rises 11 p.m., highest 3 a.m., sets in daylight.
1 July: Rises in daylight, highest in daylight, sets 1 a.m.
1 October: Rises in daylight, highest in daylight, sets in daylight.

Arcturus in Boötes is easily recognized because of its brilliance and its light orange colour; it is shown on Key Map I. Corona Borealis is unmistakable because of its shape. The other groups, apart from Scorpius, are less easy to identify, because they cover large areas of the sky and contain few bright stars.

Boötes.   Chief stars: Alpha (Arcturus) (-0.04), Epsilon (2.4), Eta (2.7), Gamma (3.0), and Delta and Beta (each 3.5). The best way to make sure of identification is to look for the large Y made up of Arcturus, Epsilon and Gamma Boötis, and Alpha Coronæ.

Doubles. Epsilon is a fine double, easy to separate, with an orange primary and a slightly bluish companion. Zeta has almost equal components, and is a rather close binary; there is an 11th-magnitude companion at a distance of 99 seconds of arc. Kappa and Xi are other easy pairs.

Variables. The brightest of the Mira stars in Boötes, R, lies close to Epsilon. In the same low-power field with Epsilon is a semi-regular variable, W Boötis, which is obviously red, and is always within binocular range.

In April 1860 J. Baxendell, a well-known variable star observer, noted a star of magnitude 9.7 in the same low-power field with Arcturus. It gradually faded, and within a few weeks had disappeared. It has never been seen again; it has been given an official designation – T Boötis – and it is worth keeping a watch, just in case it turns out to be a recurrent nova. Modern photographs show no star as bright as magnitude 17 in the position given by Baxendell.

Clusters and nebulæ. There are few nebular objects in Boötes, but the globular cluster NGC 5466 is fairly easy to find. It presents no special features.

Corona Borealis.   This beautiful little constellation cannot be mistaken, and it really does look rather like a crown. The chief stars are Alpha (Alphekka) (2.3) and Beta (3.7). The other stars in the semi-circlet are Gamma (3.8), Delta (4.6) and Theta (4.1). Despite its small size, Corona is exceptionally rich in interesting objects.

Doubles. Zeta is a fine pair, with rather unequal components; a very small telescope will divide it. Sigma is also easy. Eta, whose main components are much closer (1 second of arc) has two fainter companions at 58 and 215 seconds of arc, respectively.

Variables. Corona offers a wide choice! R Coronæ is the prototype of the 'sooty' stars; T, the recurrent nova, flared up in 1866 and 1946; S is a Mira star, and U an Algol-type eclipsing binary. All these will repay attention, particularly the first two.

Hercules.   A very large but rather barren constellation. The chief stars are Beta and Zeta (each 2.8), Alpha (Rasalgethi) (3.0 at maximum), Delta (3.1), Pi (3.2), Mu (3.4) and Eta (3.5). Hercules occupies the region between Vega and Corona Borealis.

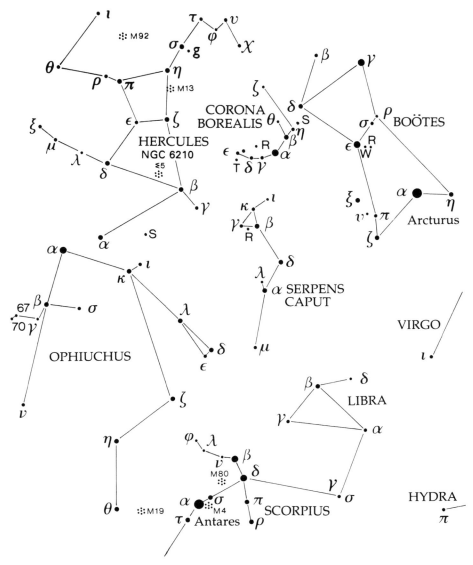

Map IX. *Boötes, Corona Borealis, Hercules, Serpens, Ophiuchus, Libra, Scorpius.*

Doubles. There are two notable pairs. Alpha, the red variable, has a companion which looks greenish by contrast, and is easily divided with a 3-inch telescope; it is worth looking at. It is a binary, but since the period has been computed as 3600 years there is little change in position angle or separation. Zeta, on the other hand, has a period of only $34\frac{1}{2}$ years, so that the changes in appearance are relatively quick. The separation is of the order of between 1 and 2 seconds of arc, and since the magnitudes are 2.9 and 5.5 the pair is not hard to split.

Variables. Alpha or Rasalgethi is a semi-regular, with a range of between 3 and 4 and a very rough period of around 100 days; I have seldom seen it fainter than 3.5. The best comparison star is Kappa Ophiuchi (3.2). Rasalgethi lies well away from the main part of the constellation, close to the brighter Alpha Ophiuchi (Rasalhague). There are several fairly bright Mira stars, and also the RV Tauri variable AC Herculis, which never fades to as low as the tenth magnitude.

Clusters. Hercules contains two splendid globular clusters, M.13 and M.92. M.13 is the finest globular visible from Britain, and is surpassed only by Omega Centauri and 47 Tucanæ in the far south; it is a superb sight in a telescope, and my 12½-inch reflector will resolve most of it. M.92 is hardly inferior, though just below naked-eye visibility so far as I am concerned; binoculars show it easily, and, like M.13, it is not difficult to resolve except close to its centre. NGC 6210 is a small but fairly bright planetary nebula lying inside the triangle formed by Beta, Delta and Epsilon Herculis.

Ophiuchus.   A very large but somewhat formless constellation; it intrudes into the Zodiac between Scorpius and Sagittarius, so that planets may pass through it. The chief stars are Alpha (Rasalhague) (2.1), Eta (2.4), Zeta (2.6), Delta (2.7), Beta (2.8), Kappa and Epsilon (each 3.2) and Theta and Nu (each 3.3). It divides the two sections of Serpens, Caput and Cauda.

Doubles. Eta is an interesting close pair with components unequal by only half a magnitude; it is a binary, with a period of just over 84 years. Rho is an easy double, lying near a very rich region.

Variables. The one really notable variable is RS Ophiuchi, a recurrent nova which reached magnitude 5 in 1933, 1958 and 1967. Apart from T Coronæ, it is the only recurrent nova which has been known to attain naked-eye visibility, and it is always worth watching, though its normal magnitude is slightly below 12.

Clusters. Ophiuchus is rich in globular clusters; among them are M.107, 12, 10, 9, 14, 62 and 19. All these are bright enough to be found without much trouble. M.12 is much less concentrated than the average globular, so that it is easy to resolve. M.62 is also unusual in as much as it is not symmetrical; the most condensed portion is well to the south-east of the centre, so that it has been likened in appearance to a comet. It is not easy to study from Britain, because it is so far south.

Serpens.   Caput is shown here (for Cauda, see Map VIII). The chief stars of Caput are Alpha (Unukalhai) (2.6) and Mu (3.5). The actual head, made up of Beta (3.7), Gamma (3.8) and Kappa (4.1) is quite distinctive.

Doubles   Delta is an easy, rather unequal pair; Beta has a companion of magnitude 9.9 at 30.6 seconds of arc. Of course the most imposing double in Serpens, Theta (Alya), lies in Cauda.

Variable. Of several Mira stars, the most important is R, easy to locate because it lies between Beta and Gamma in the Serpent's head. At maximum it is an easy naked-eye object, but it becomes very faint at minimum, and you will need good charts to identify it. The spectrum is the type M, and the star is very red.

Clusters. M.5, some way off a line joining Alpha Serpentis to Beta Libræ, is one of the finest globulars in the sky; it is not far below naked-eye visibility, and is very easy to resolve. It may be said that it is the most spectacular globular cluster visible from Britain apart from M.13 in Hercules. Its distance from us is about 27 000 light-years, and it was discovered by Gottfried Kirch as long ago as the year 1702.

Libra.   Zodiacal, but frankly a rather dull constellation. The chief stars are Beta (2.6), Alpha (2.7) and Sigma (3.3); Sigma was formerly included in Scorpius, as Gamma Scorpii. Beta Libræ is said to be the only single star with a decidedly greenish hue, though I admit that I have never seen it as anything but white; the spectral type is B8.

Double. Alpha is a very wide double, well seen with binoculars.

Variables. There are several Mira stars which can attain about the seventh magnitude, and also the Algol-type eclipsing binary Delta Libræ, which is in the same binocular field with Beta.

Scorpius.   A splendid Zodiacal constellation, but always inconveniently low as seen from England; the southernmost part of it does not rise, and it is never easy to make out the magnificent 'sting', including one star (Shaula) not much below the first magnitude. The chief stars of Scorpius are Alpha (Antares) (1.0), Lambda (Shaula) (1.6), Theta (Sargas) (1.9), Epsilon and Delta (each 2.3), Kappa (2.4), Beta (2.6), Upsilon (2.7), Tau (2.8), Sigma and Pi (each 2.9), Iota^1 and Mu (each 3.0), G (3.2), and Eta (3.3). The long chain of bright stars really does conjure up some impression of a scorpion! The 'head' (Beta, Nu and Omega) is distinctive; Antares, the brilliant red supergiant, is flanked to either side by Sigma and Tau. Scorpius is very rich, and abounds in interesting objects.

Doubles. Antares has a greenish companion which can be seen with a 3-inch refractor; the companion is known to be a radio source. Beta is very wide and easy; the brighter star is itself a close double. Both Zeta and Mu give the impression of being naked-eye doubles, but in the case of Zeta there is no real association; the fainter member of the pair is much the more luminous, and is much further away.

Variables. There are various Mira stars here, though none can lay claim to being particularly notable.

Clusters and nebulæ. There are several bright open clusters, including M.6 (the 'Butterfly') and M.7, both of which are visible with the naked eye and are superb in binoculars or with a low power. NGC 6124 is another imposing open cluster. There are two fine globulars, M.4 (near Antares) and M.80 (near Delta). M.4 is not far inferior to M.13 Herculis. M.80 was the site of a bright nova (T Scorpii) seen in 1860 near the centre of the cluster; it reached magnitude 7, and quite outshone the cluster itself. It slowly faded, and was lost when it dropped below magnitude 12. Unless it were a recurrent nova, it will not be seen again, but perhaps M.80 is worth checking whenever possible. It is not too easy to resolve except in its very outermost parts.

Map X. Pegasus, Andromeda, Pisces, Triangulum, Aries, Cetus, Aquarius, Sculptor, Piscis Australis.

The chief constellation in this map is Pegasus, whose four main stars form the famous Square. One way to identify it is by means of a line from Gamma Cassiopeiæ through

Alpha. The line from Merak and Dubhe through Polaris will also reach the Square if prolonged far enough across the sky. Very rough rising and setting times are:

1 January: Rises in daylight, highest in daylight, sets in daylight
1 April: Rises 2 a.m., highest in daylight, sets in daylight.
1 July: Rises in daylight, highest 5 a.m., sets in daylight.
1 October: Rises in daylight, highest 11 p.m., sets 7 a.m..

It is therefore at its best during autumn. As is shown on Map VII, one of the stars of the Square, formerly Delta Pegasi, is now included in Andromeda, as Alpha Andromedæ. Andromeda and Triangulum have been described with Map VII.

Pegasus.   An important constellation, though not so conspicuous as might be imagined from the map; the Square is large, and its stars are not particularly bright. The leaders of Pegasus are Epsilon, Enif (not in the Square) (2.4), Alpha (2.5), Beta (Scheat) (variable, 2.4 at maximum), Gamma (2.8), Eta (2.9), Zeta (3.4) and Mu and Eta (each 3.5). It is interesting to see how many stars you can count inside the Square without using optical aid. Try it, and you may have quite a surprise.!

Variables. Beta is a semi-regular variable, easy to follow by comparing it with Alpha and Gamma – but take care not to forget extinction. The rough period is 38 days, and the usual range is between 2.4 and 2.8. R Pegasi is a normal Mira star.

Cluster. M.15 is a bright globular, near Epsilon, though actually closer to Delta Equulei. It has a small, very bright centre, and is surrounded by a large area rich in faint stars. It contains a planetary nebula of magnitude 13.8 which is, however, a very difficult object indeed.

Aries.   This small but quite distinctive constellation is offiically regarded as the first in the Zodiac, even though the 'first Point of Aries' has now shifted into Pisces by virtue of precession. The leaders of Aries are Alpha (2.0) and Beta (2.6); the third member of the trio is Gamma (combined magnitude 3.9). Alpha, or Hamal, is an orange K-type star. The only really notable object is Gamma or Mesartim, which is a fine double; the components are equal, and since the separation is almost 8 seconds of arc the pair is extremely easy to split even in a small telescope.

Pisces.   A dim Zodiacal constellation, sprawling along south of the Square of Pegasus. Its brightest star is Eta (3.6).

Doubles. Zeta is wide and easy. Alpha is much closer, with a separation of below 2 seconds of arc, but it is certainly not difficult, and the difference between the two components is less than a magnitude.

Variables. The most interesting variable is TX (19) Piscium, close to the little group made up of Lambda, Iota, Theta and Gamma. It has an N-type spectrum, and is intensely red; I make it almost as vivid as Mu Cephei. It varies irregularly, and remains rather below naked-eye visibility, but binoculars will always show it.

Galaxy. M.74 is an open spiral not far from Eta, but it is not an easy object, and may be regarded as the most elusive of all the Messier objects. Neither will it bear a high magnification. The best chance of finding it is to use a very low power.

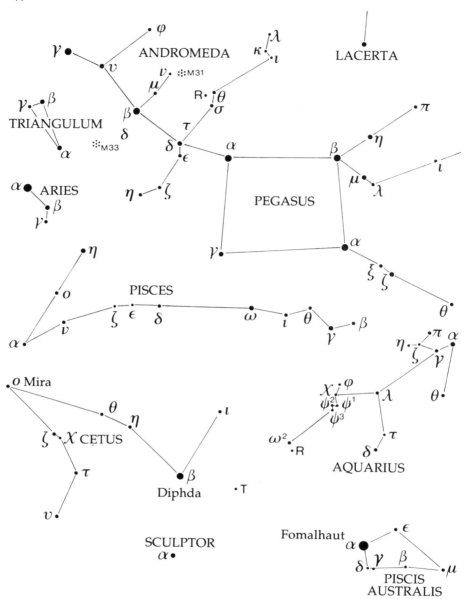

Map X. Pegasus, Andromeda, Pisces, Triangulum, Aries, Cetus, Aquarius, Sculptor, Piscis Australis.

Cetus.   Part of Cetus is shown in Map IV, and has been described there. The chief stars shown in the present map are Beta (Diphda) (2.0) and Eta and Tau (each 3.5). Diphda can be found from the Square of Pegasus, since Alpha Andromedæ and Gamma Pegasi point to it. It is suspected of variability. Not far from it is the red semi-regular variable T Ceti. Mira, just on this map, has been described with Map IV.

Sculptor. A very dim constellation adjoining Cetus; the brightest star is Alpha (4.3). Apart from a couple of red variables, R and S, there is little of interest here for owners of small telescopes, though more powerful instruments will show a number of faint galaxies, and there is one – NGC 7793 – which is fairly prominent; it lies almost edgewise-on to us.

Aquarius. Part of this Zodiacal constellation is shown in Map VIII, but most of it is in the present map. The chief stars are Beta (2.9), Alpha (3.0), Delta (3.3) and Zeta (3.6). Aquarius is rather a formless group, lying between Pegasus and Pisces to the north and Fomalhaut to the south.

Double. Zeta is a fine slow binary, with equal components, very easy to separate. Beta has a companion of magnitude 10.8 at a distance of 35.4 seconds of arc.

Variables. Aquarius contains several unremarkable Mira and semi-regular stars, but the most interesting variable is the peculiar R Aquarii, which is of the 'symbiotic' type and is never predictable. It is a binary, and is associated with nebulosity. Amateurs will do well to keep a watchful eye upon it.

Clusters and nebulæ. Both the Aquarius globulars, M.2 and M.72, lie on Map VIII; so does the Saturn Nebula, a planetary. The other important planetary in Aquarius is NGC 7293, the Helix, which is large but has low surface brightness; it is best found with a low power, and it will not stand magnification nearly as well as some other planetaries. Photography is needed to show it really well.

Piscis Australis (or Piscis Austrinus) This is distinguished by the presence of Alpha (Fomalhaut), which is of magnitude 1.2, but there are no other stars above the fourth magnitude. Fomalhaut is always very low as seen from Britain, and observers in North Scotland will be lucky to see it at all, but during autumn evenings it rises briefly; look for it below the Square of Pegasus, using Beta and Alpha as direction-finders. Fomalhaut is one of the stars found by IRAS, the infra-red satellite, to be associated with cool, possibly planet-forming material; at its distance of 22 light-years it is one of our nearer neighbours, and it is 13-times as luminous as the Sun. There is not much else of interest in Piscis Australis, though Beta is an easy double – not a binary, but an optical pair.

THE SOUTHERN-HEMISPHERE VIEW

Map XI. Key Map: Crux Australis (the Southern Cross).

Map I–X have been drawn for observers who live in the northern hemisphere. In fact the maps are valid for latitudes well south of Europe, with certain modifications; they do not, of course, apply to countries such as South Africa, Australia or New Zealand. Moreover, the orientation is reversed, so that, for example, the stars of Orion's Belt point upward to Sirius and downward to Aldebaran, while Leo and Virgo appear as shown in the map below.

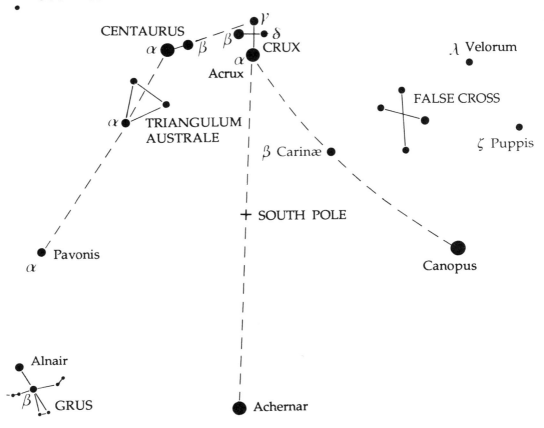

Map XI. Key Map: Crux Australis (the Southern Cross).

For this book I have not thought it necessary to re-draw all the maps, because it is easy to re-orientate them. However, we have not yet discussed the stars which are too close to the south pole to be seen from Europe or most of the United States. We cannot use the Great Bear as a pointer, but instead we have the Southern Cross, and also Orion, which is cut by the celestial equator and can therefore be seen from every inhabited country in the world.

The Southern Cross is the smallest constellation in the sky, but it is also one of the most striking. Let us admit that it is not in the least like a cross; it is more in the shape of a kite. Of its four chief stars, three are brilliant; the fourth, Delta Crucis, is much fainter, and spoils the symmetry of the pattern.

There should be no problem in identifying Crux, particularly as the brilliant Alpha and Beta Centauri show the way to it; like Dubhe and Merak in the Great Bear, they are often nicknamed the Pointers. In Crux, Alpha (Acrux) and Beta are of the first magnitude, and Gamma only just below.

From South Africa, Crux can drop very low in the sky, as on spring evenings (that is to say, around November). From Johannesburg part of it actually sets, though from places further south, such as Cape Town, it scrapes the horizon, and from much of New Zealand it is circumpolar. The maps given in the rest of this section may be regarded as valid for the whole of South Africa, Australia and New Zealand, though again with some modifications.

The first thing to learn from Crux is the position of the south celestial pole, which lies in a decidedly blank area. Simply follow the 'longer axis' of the Cross, from the red giant Gamma through Acrux. After passing through the polar region, the line will come to Achernar, the brilliant leader of Eridanus. Achernar and Crux are on opposite sides of the pole, and about the same distance from it — so that when Achernar is high up, Crux will be low down, and vice versa.

Next, follow the 'sweep' shown in Map XI. Beta Carinæ in the Keel of the old Ship Argo is of magnitude 1.7, and is therefore bright enough to be conspicuous; beyond it we come to Canopus, which is surpassed only by Sirius. Sirius itself lies well beyond Canopus, too far to be conveniently shown in this key map. Beware of the False Cross, which lies partly in Carina and partly in Vela. In shape it is very like Crux, but it is much larger, and its stars are not so bright.

Close to Alpha Centauri, the brighter of the Pointers, is Triangulum Australe, the Southern Triangle — one of the few groups to have been given an appropriate name. Alpha, the brightest of the trio, is of the second magnitude, and is distinctly reddish. Follow a line from Alpha Centauri through Alpha Trianguli Australe, and you will eventually come to the second-magnitude star Alpha Pavonis. Beyond lies Grus, the Crane, with its second-magnitude leader Alnair. I have given these stars in the key map because they are the most easily identified objects in an area which is rather confused and undistinctive apart from Grus itself. Alpha Pavonis is circumpolar from Australia, but Grus is not, and during winter evenings it drops below the horizon.

Various other constellations can be found from Crux. For instance, a line from Acrux through Gamma will show the way to Corvus, so that to all intents and purposes Achernar, the south pole, Crux and Corvus are lined up. Acrux and Alpha Centauri act as approximate guides to Antares, and a line from Acrux passed midway between Beta and Gamma Crucis will end up somewhere near Spica in Virgo.

Map XII. Key Map: Orion.

Orion is on view for a large part of the year — all through the hot season — and is out of sight only during the winter. Canopus can be located by using Zeta and Kappa Orionis, and Canopus and Sirius point to the Twins, Castor and Pollux. Regulus and the Sickle of Leo can be found by taking a 'sweep' from the lower part of Orion through Procyon, as shown in the key map. Canopus is not circumpolar from South Africa or most of Australia, but it spends little time below the horizon.

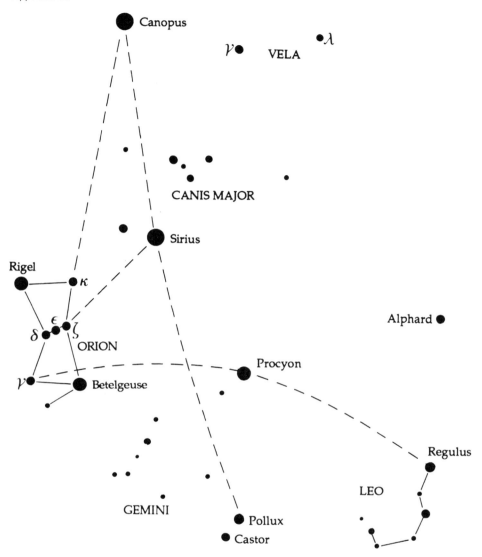

Map XII. Key Map: Orion.

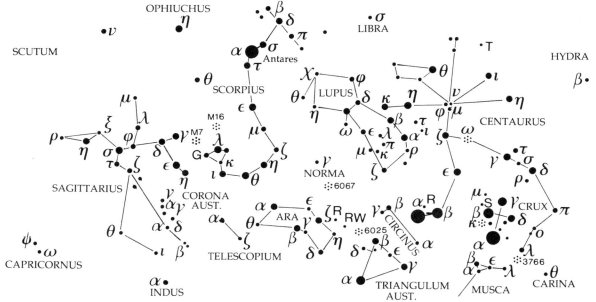

Map XIII. Crux, Centaurus, Lupus, Circinus, Triangulum Australe, Norma, Ara, Telescopium, Sagittarius, Corona Australis, Scorpius.

Map XIII. Crux, Centaurus, Lupus, Circinus, Triangulum Australe, Norma, Ara, Telescopium, Sagittarius, Corona Australis, Scorpius.

This map covers a brilliant region of the sky, crossed by the Milky Way. In addition to the Crux–Centaurus area, it includes the whole of Sagittarius and most of Scorpius.

Crux. Crux is pre-eminent. Its leading stars are Alpha (Acrux) (0.8), Beta (1.2), Gamma (1.6) and Delta (2.8); then comes Epsilon (3.6). Beta has no official proper name, though it is sometimes referred to as Mimosa. Though there is only 0.4 magnitude difference between Beta and Gamma, Beta is always ranked as a first-magnitude star, while Gamma is not; note also that Gamma is red, and of type M, whereas the other three leaders are hot white stars.

Doubles. Acrux is one of the most imposing doubles in the sky. Any small telescope will show both components, with the third star in the same field. Gamma has a 7th-magnitude companion, and also a much fainter companion (9.5) at a distance of 115 seconds of arc. Theta^1 and Theta^2 (4.3 and 4.7) give the impression of being a naked-eye double, but they are not related, since the fainter star is much the more luminous and remote; Theta^1, the brighter member, has a very faint companion. Mu^1 is a wide, very easy telescopic double.

Variables. None of these can be regarded as very distinguished, but R, S and T are all Cepheids which remain within binocular range.

Clusters and nebulæ. There are several open clusters in Crux, but pride of place must go to NGC 4755 – Kappa Crucis, nicknamed the Jewel Box. It is often said that the name has been given because there are stars of contrasting colours; actually most of them are bluish-white, but there is one red leader which stands out magnificently. The Jewel Box is best seen with a low or moderate power, and ranks with the finest objects

319

of its type. There is also the dark nebula which is known as the Coal Sack, abutting on Alpha and Beta; it is easy to make out a starless area, with the naked eye, and even in a telescope there are not many field stars in front of it. It covers an area of 26.2 square degrees, and is undoubtedly much the most striking of all the dark nebulæ.

Centaurus. A truly grand constellation, more or less surrounding Crux; indeed, until near-modern times Crux was included in it. The chief stars of Centaurus are Alpha (− 0.3), Beta (Agena) (0.6), Theta (2.1), Gamma (2.2), Epsilon and Eta (each 2.3), Zeta (2.5), Delta (2.6), Iota (2.8), Mu (2.9 at maximum), Kappa and Lambda (each 3.1) and Nu (3.4). As we have noted, Alpha has no official proper name, though Toliman is one suggestion, and some air and sea navigators refer to it as Rigel Kent. Beta has an alternative proper name: Hadar.

Variables. There are several Mira and semi-regular stars in Centaurus, notably R, which lies roughly between Alpha and Beta and can rise to magnitude 5.3. S Centauri has an N-type spectrum, and is very red. Mu is a curious sort of variable; it has a B-type spectrum, and seems to fluctuate irregularly between magnitudes 2.9 and 3.5, but it will repay attention. Iota and Nu are suitable comparison stars.

Doubles. Alpha Centauri is one of the finest binaries in the sky, separable with any small telescope; it has a period of just under 80 years. Gamma is also a binary, with equal components, but the separation is only about 1.4 seconds of arc. Beta has a fourth-magnitude companion at 1.3 seconds of arc, which is not too easy because it is so overpowered by the brilliance of the primary.

Clusters and nebulæ. Omega Centauri, easily found because it lies in line with Beta and Epsilon, is much the brightest globular cluster in the sky, and is prominent with the naked eye as hazy patch. Yet surprisingly, in a telescope it is probably less impressive than 47 Tucanæ, because it more than fills the field unless a very low power is used. There are various other open clusters, notably IC 2944, which is associated with Lambda Centauri. The galaxy NGC 5128 is a fine sight; it is irregular in form – it was once, wrongly, though to be made up of two galaxies in collision – and is a strong radio source, so that it is also known as Centaurus A. To see it properly requires a fairly high magnification, but small telescopes will show it as a hazy patch.

Circinus. A small constellation near Alpha and Beta Centauri. The only star above the fourth magnitude is Alpha (3.2) and there are few objects of note, though Gamma is a binary with fairly equal components and small separation.

Norma. Another small constellation, with no star brighter than Gamma² (4.0). The only object of note is the open cluster NGC 6087, which lies between Alpha Centauri and Zeta Aræ; it is fairly rich, and contains a Cepheid variable, S Normæ, which has a range of from 6.1 to 6.8 and a period of 3.75 days.

Triangulum Australe. Quite prominent; the triangle consists of the reddish Alpha (1.9), Beta (2.8) and Gamma (2.9). There is however not much of interest here, apart from a fairly bright and easily resolvable open cluster, NGC 6025. S Trianguli is a typical Cepheid, well within binocular range.

Ara. A fairly distinctive constellation, between Theta Scorpii on one side and Triangulum Australe on the other. The chief stars are Beta (2.8), Alpha (2.9), Zeta (3.1) and Gamma (3.3); Zeta is reddish, and of type K.

Variable. R Aræ is in the same low-power field with Zeta. It is an Algol-type eclipsing binary, always remaining above the seventh magnitude.

Clusters and nebulæ There are several open clusters, and three fairly bright globulars, NGC 6352, 6362 and 6397. Of these NGC 6397 is notable as being only about 8200 light-years away, in which case it is the closest of all the globulars. It is not, however, very condensed.

Telescopium. A small constellation adjoining Ara; the chief stars are Alpha (3.5) and Zeta (4.1). The main object of interest is the Z Andromedæ variable RR Telescopii, once regarded as a curious type of nova, not far from Alpha Pavonis. Originally it seems to have been a variable with a range of from 12.5 to 15 and a period of 387 days; subsequently the minima became deeper (magnitude 16) and then, in 1944, the magnitude suddenly increased to 7, rising to 6.5 in 1949. It then slowly faded. It is not particularly easy to identify, and detailed charts are needed, but it is unquestionably worth watching.

Lupus. A rather formless constellation; The brightest stars are Alpha (2.3), Beta (2.7) and Gamma (2.8).

Corona Australis (or Corona Austrinus). This is by no means so prominent as its northern namesake; it lies near the strangely-obscure Alpha and Beta Sagittarii, and is easy to identify from its shape, though it has no star as bright as the fourth magnitude. There are two easy doubles (Kappa and Gamma) and a fairly prominent globular cluster (NGC 6541). There is also the dim nebula NGC 6729, containing the variable R Coronæ Australis, which however never rises above magnitude 9.7; at minimum it falls to about 12.

Map XIV. Carina, Vela, Puppis, Volans, Pyxis, Antlia.

This is the region of the vast constellation Argo Navis, now divided up into a keel, sails and a poop. Crux, Beta, Carinæ, Canopus and Sirius form a magnificent curved line, which cannot be mistaken. The False Cross is made up of Epsilon and Iota Carinæ, and Kappa and Delta Velorum. The whole area is exceptionally rich, and is crossed by the Milky Way.

Carina. This is the brightest part of Argo, and is dominated by Canopus, inferior only to Sirius in the whole sky – and very much more powerful; according to the

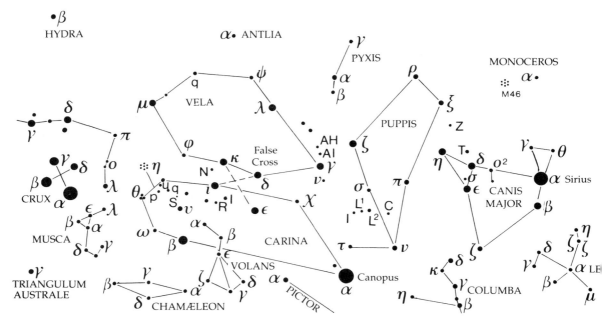

Map XIV. Carina, Vela, Puppis, Pyxis, Volans, Antlia.

Cambridge Catalogue it has a luminosity 200000 times that of the Sun. It has an F-type spectrum, and should therefore be slightly yellowish, but neither with the naked eye nor with optical aid have I ever been able to see it as anything but white. The leading stars of Carina are Alpha (Canopus) (− 0.7), Beta (Miaplacidus) (1.7), Epsilon (1.9), Iota (2.2), Theta (2.8), Upsilon (3.0), p and Omega (each 3.3), q and a (3.4) and Chi (3.5). Volans intrudes into Carina between Canopus and Miaplacidus.

Variables. There are several variables bright enough to be seen with the naked eye at maximum, including the Mira stars R and S, and ZZ (or l) Carinæ, which is a Cepheid with the unusually long period of 35.5 days and a range of from 3.3 to 4.2. However, the most fascinating object by far is Eta Carinæ. It has already been described in the text; it was for a while, more than a century ago, the brightest star in the sky apart from Sirius, but for many years now it has been on the fringe of naked-eye visibility. It looks quite unlike a normal star, and is orange in colour; it is associated with the nebula NGC 3372, which includes the dark Keyhole Nebula. I recommend looking at it with a moderate magnification, as there is nothing else like it.

Clusters and nebulæ. There are many interesting nebular objects in Carina, quite apart from NGC 3372. The open cluster IC 2602, round Theta, is very prominent with the naked eye, and is believed to be extremely young; it is well seen with binoculars or a low-power eyepiece. NGC 2516 is another glorious open cluster, in line with Kappa Velorum and Epsilon Carinæ in the False Cross. There is also a bright planetary, NGC 2867, between Iota Carinæ and Kappa Velorum.

Vela. The sails of the old Ship. The chief stars are Gamma (1.8), Delta (2.0), Lambda (2.1), Kappa (2.5), Mu (2.6), N (3.1) and Phi (3.5). N, a K-type reddish star, has been suspected of variability, but is more probably constant in brightness. Gamma

Velorum (Regor) is the brightest of the W-type stars, and is very hot and unstable.

Doubles. Gamma is a fine, easy double; there are three more distant companions. Delta is also an easy pair, with two faint companions; Mu is a binary, with a period of over 100 years.

Variables. There are several reasonably bright Cepheids and Algol stars in Vela; the brightest Mira variable is Z, which reached magnitude 7.8 at maximum but becomes very faint indeed near minimum, so that a telescope of considerable aperture is needed to follow it. Also of interst is AI Velorum, a Delta Scuti variable with a range of from 6.4 to 7.1 and a period of less than three hours.

Clusters and nebulæ. Vela contains several bright open clusters, notably IC 2391, close to Delta; the brightest star in it is o Velorum (3.6), and even with binoculars it is seen to have a vaguely cruciform appearance. NGC 2547, near Gamma, is also visible with the naked eye. The planetary nebula NGC 3132, on the borders of Vela and Antlia, has been compared with the Ring Nebula M.57 Lyra, but has a much brighter central star. Its actual diameter is about half a light-year, and it is decidedly elliptical in form.

Volans. This constellation intrudes into Carina, and has no star brighter than Gamma (3.6); this is actually a combined magnitude, since Gamma is a wide, easy double. There is nothing else of immediate interest in Volans.

Antlia. An obscure constellation adjoining Vela; the brightest star, Alpha, is only of magnitude 4.2. Theta is a close double with almost equal components. U Antliæ is a red N-type variable; periods of from 170 to as much as 365 days have been assigned to it, though in other catalogues it is classed as irregular. It could be well worth some attention.

Puppis. The Poop of the Ship. Chief stars: Zeta (2.2), Pi (2.7), Rho (2.8), Tau (2.9), Nu and Sigma (each 3.2), Xi (3.3) and L^2 (3.4 at maximum). Part of Puppis northerly enough to rise in the latitudes of Britian.

Variables. L^2 is an M-type semi-regular star, which can reach magnitude 2.6 and never falls below 6.2; it is therefore a naked-eye object for most of the time, and is always within binocular range. V Puppis is a good example of a Beta Lyræ-eclipsing binary, but it has a small range of only half a magnitude.

Clusters and nebulæ All three Messier objects in Puppis – 46, 47 and 93 – are in the northern part of the constellation, and can be seen from British latitudes, though they are always low down. There is also the fine cluster NGC 2477, near Zeta, which is smaller, richer and more compact than M.46. The whole of Puppis is rich, and well worth sweeping with a low magnification.

Pyxis. A small constellation in the Vela–Puppis area. Its brightest star is Alpha (3.7). The only object of real note is the recurrent nova T Pyxidis, which is normally of magnitude 14, but may rise unpredictably to above 7 – as it did in 1890, 1902, 1920,

1944 and 1966. Obviously it is worth keeping under watch, as another outburst may occur at any moment. Its increases are rather less abrupt than with other recurrent novæ, such as T Coronæ Borealis and WZ Sagittæ.

Map XV. Octans, Apus, Musca, Chamæleon, Mensa, Hydrus, Reticulum, Dorado, Pictor, Eridanus, Horologium.

This is the south polar area. It is divided up into a number of relatively dim constellations, and it is not too easy to find one's way about, particularly since when the sky is not completely dark the whole region tends to appear blank. Broadly speaking, it is enclosed by imaginary lines joining Canopus, Achernar, Alpha Pavonis and Alpha Trianguli Australe. As has been shown on the key map, the pole may be located by using the longer axis of Crux as a guide.

Octans. This contains the south pole; the brightest star is Nu (3.8). Sigma Octantis, magnitude 5.5, is the actual south pole star, marking it to within a degree, but it is a very poor substitute for the northern hemisphere's Polaris. I have worked out a good way of finding it, but binoculars are usually needed.* There is little of interest in Octans; R is a fairly bright Mira star, and there are a couple of undistinguished doubles, Iota and Lambda.

Apus. A fairly compact little group; the brightest star is Alpha (3.8). Delta is a wide double, and there are two variables of interest; Theta, which lies close to Alpha, and R Coronæ-type star S Apodis, which is however faint even at maximum, and becomes very difficult at minimum, so that to follow it means using a fairly large telescope together with a set of detailed charts.

Musca. A small but distinctive constellation close to Crux. The chief stars are Alpha (2.7), Beta (3.0) and Delta and Lambda (each 3.6). Lambda and Mu (4.7) make up a striking pair, in the same low-power binocular field with Acrux; Lambda is white, Mu very red. Beta Muscæ is a fairly easy double, and R is a typcial Cepheid, but the most interesting objects in the constellation are the two globular clusters NGC 4372 and 4833. Both are fairly bright, and not hard to resolve.

Chamæleon. Again we have a faint, obscure constellation, with its main stars making up a sort of diamond shape, but all are below the fourth magnitude. R is a Mira variable and both Delta and Epsilon are close doubles, but on the whole Chamæleon contains little of note.

Mensa (originally Mons Mensæ, the Table Mountain). This seems to be an entirely unnecessary constellation, with no star above the fifth magnitude and no notable objects – though it is true that a small part of the Large Magellanic Cloud extends into it from neighbouring Dorado.

* See *Exploring the Night Sky with Binoculars* (Cambridge 1986, p.121).

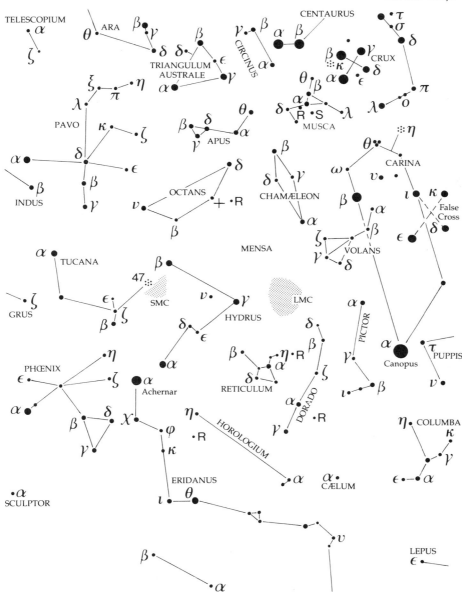

Map XV. Octans, Apus, Musca, Chamæleon, Mensa, Hydrus, Reticulum, Dorado, Pictor, Eridanus, Horologium.

Hydrus.   The main stars are Beta (2.8), Alpha (2.9) and Gamma (3.2). Beta Hydri is the nearest fairly bright star to the south celestial pole, though it is well over 10 degrees away. Apart from the SS Cygni variable VW Hydri, which never becomes as bright as the eighth magnitude, Hydrus is singularly lacking in objects of interest.

Reticulum.   This is easy to identify because of its compactness; the chief stars are Alpha (3.3) and Beta (3.8). There is not much of special interest, though R Reticuli is a fairly bright Mira star inasmuch as it can attain magnitude 6.5 at maximum.

Pictor.   Near Canopus; the leading star is Alpha (3.3). Beta (3.8) is of special interest; it was one of the stars found by the IRAS satellite to have a large infra-red excess, and this material has subsequently been photographed. It may well be that Beta Pictoris is the centre of a planetary system. According to the Cambridge catalogue it is 78 light-years away and at least 60 times as luminous as the Sun. There is nothing else of much note in Pictor, though the Mira variables R and S are quite bright when near maximum. A conspicuous nova, RR Pictoris, flared up here in 1925, but has now become extremely faint.

Dorado.   In itself Dorado is not particularly notable; the brightest star is Alpha (3.3). Beta Doradûs is a naked-eye Cepheid, and R is a red semi-regular variable always within binocular range. However, Dorado includes most of the Large Magellanic Cloud (LMC), which is only 170 000 light-years away – and is much the brightest of the external galaxies; it includes the magnificent Tarantula Nebula NGC 2070 (30 Doradûs). It is a superb sight in any small telescope. It was here that a bright supernova flared out in 1987, and became very conspicuous with the naked eye, remaining visible without optical aid for some months.

Eridanus.   Part of this very long constellation is shown here; the brightest stars in this section are Alpha (Achernar) (0.5) and Theta (Acamar) (2.9). Theta is a splendid double, easily separated in a small telescope. In ancient times it was ranked as of the first magnitude, so that either it has faded in the meantime or else (more probably) there is some confusion in the old records between it and Achernar. Achernar itself, one of the brightest stars in the sky, is not too far from the south pole. It lies close to Alpha Hydri.

Horologium.   An obscure constellation between Eridanus and Reticulum/Dorado; its brightest star is Alpha (3.9). There is one Mira star, R Horologii, which can reach magnitude 4.7 at maximum, though at minimum it drops below magnitude 14; there is one fairly bright globular cluster, NGC 1261, and a number of galaxies, most of which are rather faint.

Also included in this map are Phœnix, Tucana and Pavo, but these are best described with Map XVI. Note that the Small Magellanic Cloud adjoins Hydrus and actually extends into it, but most of it lies in Tucana.

Map XVI. Pavo, Indus, Tucana, Grus, Phœnix, Microscopium.

This is the region of the 'Southern Birds'. I have found that the best method of identification is to locate Alpha Pavonis by using Alpha Centauri and Alpha Trianguli Australe, as shown in the key map. Of all these groups, only Grus is really distinctive.

Pavo.   The chief stars are Alpha (1.9), Beta (3.4) and Delta and Eta (each 3.6). Alpha is some way away from the main part of the constellation; Delta, at a distance of 19 light-years, is almost exactly similar to the Sun in size, mass and luminosity.

Variables. There are two naked-eye variables in Pavo. Kappa is a Type II Cepheid, which means that it is much less luminous than a classical Cepheid of the same period; also, it is rather less precise in its behaviour. Lambda is also variable to the extent of about a magnitude, but has a B-type spectrum, and seems to be irregular.

Clusters and nebulæ. Pavo contains one exceptionally fine globular cluster, NGC 6752, which is not greatly inferior to M.13 in Hercules; it is moderately condensed, and at a distance of around 20000 light-years it is one of the nearer globulars. There are also several galaxies within the range of telescopes of moderate aperture.

Indus. A small constellation adjoining Pavo; the chief star is Alpha (3.1). Theta is an easy double, and Delta a much closer one with equal components. There are two red variables, one Mira star (S Indi, type M) and one semi-regular (T Indi, type N). There are also some rather faint galaxies.

Tucana.   The least distinctive of the Southern Birds; the brightest star is Alpha (2.9). Alpha Tucanæ, Alpha Pavonis, and Alnair (Alpha Gruis) form a triangle.

Doubles. Beta is a fine multiple star. The main pair is made up of two components, Beta¹ (4.4) and Beta² (4.8), giving a combined magnitude of 3.7. Beta¹ has a 14th-magnitude companion at a distance of 2.4 seconds of arc; Beta² is a binary, with components of magnitudes 4.8 and 6.0 and a period of 44.4 years. In the same low-power field is Beta³, which is a close double of equal six-magnitude components and a separation of 0.1 seconds of arc. This whole group will repay attention. Kappa Tucanæ (5.4) has a companion of just below the seventh magnitude, and Delta (4.5) has a ninth-magnitude companion at a separation of almost 7 seconds of arc.

Clusters and nebulæ.   Tucana contains almost the whole of the Small Magellanic Cloud (SMC), which, though by no means the equal of the Large Cloud, is still a prominent naked-eye object. It is slightly further away than the Large Cloud, and is possibly a compound system. It was by studying Cepheids in it that Miss Leavitt, in 1912, discovered the period–luminosity law which has been so invaluable in our distance-gauging. Almost silhouetted against the Small Cloud is the splendid globular cluster 47 Tucanæ, the brightest in the sky apart from Omega Centauri; in my view it is the more spectacular of the two when viewed through a telescope, because it does not fill the entire field. It is interesting to note that the surface brightness of 47 Tucanæ is much greater than that of the SMC; it is always tempting to assume that the two are

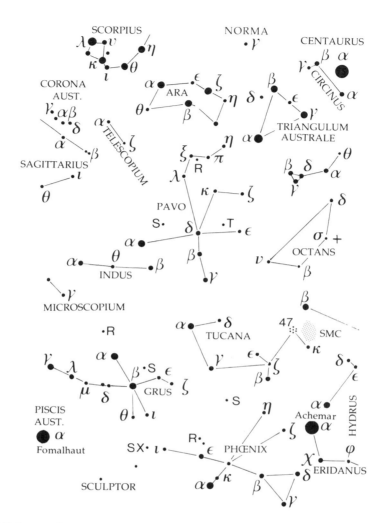

Map. XVI. Pavo, Indus, Tucana, Grus, Phœnix, Microscopium.

associated, though in fact there is no real connection between them — 47 Tucanæ belongs to our Galaxy. Also in Tucana is another fine globular, NGC 362, which is on the fringe of naked-eye visibility, and is not hard to resolve except at its centre.

Grus. The celestial Crane is very prominent, and does indeed give some impression of a bird in flight. The chief stars are Alpha (Alnair) (1.7), Beta (Al Dhanab) (2.1), Gamma (3.0) and Epsilon (3.5). The two leaders make an interesting contrast; Alnair, a white star of type B, differs markedly from the M-type giant Al Dhanab, which is a lovely warm orange in colour. Grus is marked by a curved line of stars, of which Delta and Mu give the impression of being wide naked-eye pairs. Actually there are not many interesting telescopic objects in Grus, but there are numerous rather faint galaxies.

Phœnix. Last in our admittedly brief survey of the sky we come to Phœnix, whose leading stars are Alpha (Ankaa) (2.4), Beta (3.3) and Gamma (3.4). It is not very distinctive, and is in the far south; Zeta Phœnicis lies close to Achernar. The triangle formed by Beta, Delta and Gamma is fairly easy to identify.

Doubles. Beta is a fairly easy double with nearly equal components; Eta has a companion of magnitude 11.4 at a distance of over 19 seconds of arc.

Variables. Zeta Phœnicis is an eclipsing binary of the Algol type, with a period of 1.67 days and a range of from 3.9 to 4.4; Epsilon (3.9) and Eta (4.4) are good comparisons. R is a typical Mira variable.

Appendix 29

Astronomical Photography

There are many subjects which I have not covered in this book. For example, I have said virtually nothing about radio astronomy, for the excellent reason that I an not a radio astronomer, and I have a rooted dislike of writing about anything upon which I do not feel competent to act as a guide. For the same reason I do not propose to say much about photography; I have never taken it really seriously, though I have, of course, taken pictures of the Moon, planets, star-fields and various special phenomena such as eclipses.

You can make a start even if you have nothing more than an ordinary camera, the one essential being that it must have a B shutter; time-exposures are always necessary. The procedure for photographing star trails is quite simple. Focus to infinity, point the camera at a dark, clear sky, and give a time exposure, ranging from a minute or two up to — well, anything you like, though with long exposures there is always the risk of dewing. Fast films are an advantage in most ways, though slower ones are less grainy. For black-and-white pictures, Pan-F is a useful film, but in general you can use any film faster than ISO 64 or thereabouts.

If you point the camera at the pole, you will get spectacular curved trails, and you may be lucky enough to snare a meteor or a satellite. (You may also snare an aircraft, but this is something which cannot be helped!) Colour film will be interesting — for example, it is easy to record the orange-red of Betelgeux in comparison with the pure white of Rigel.

For wide areas of the sky you naturally need a wide-angle lens; for smaller areas you can use a standard or a telephoto. Always put the camera on to a tripod, and use a cable release.

If you give an exposure of more than a minute or so, you are bound to get star trail. One way to avoid this is to use what is called a Haig mount, so named in honour of its inventor, George Haig. It is essentially a simple equatorial mounting for a camera (for full details, see the article by H.J.P. Arnold in *Astronomy Now*, Vol. 1, No. 1, p.39). Using it, you can record stars with an exposure of at least ten minutes without any appreciable trail.

Again using a fixed camera, the Moon is an interesting subject; at full, an exposure of $\frac{1}{60}$ of a second at f/11, with a 100 ISO film, will give good results, while the crescent will need about $\frac{1}{30}$ second at f/11 with 400 ISO film. Always 'bracket' exposures – that is to say, take some pictures with longer exposures than you think are really necessary, and do the same with shorter exposures. And always keep full notes, so that you can see which exposures are the most successful. Also, it is best to do your own developing and printing if possible. Take your films to the local chemist, and he may well tell you that they are useless – because they look blank. If you are using a commercial firm, ask for your negatives to be returned uncut.

Eclipses are, of course, ideal subjects for the photographer. Lunar eclipses are tranquil affairs, and you have plenty of time; with ISO 100, for example, an exposure of $\frac{1}{250}$ second at f/11 will do when the eclipse is just beginning, but at mid-totality you will need something more like 2 seconds at f/2. A total solar eclipse is much briefer, so make sure that everything has been rehearsed before hand – and remember, too, never to look at the Sun direct through the view-finder of an SLR camera. Bracket your exposures, and leave nothing to chance. I have used 250 mm and 500 mm telephoto lenses with success, but everything really depends upon what you want to do – corona, chromosphere, or a general impression.

If you have a clock-driven telescope, the camera may be mounted on it, so that the telescope is used merely as a guide. If you want to use the telescope itself, there are various methods. In prime focus photography, the camera body is plugged into the drawtube of the telescope, and used as a very long telephoto lens. In eyepiece projection, the camera is extended out from the eye-piece, either by means of bellows or extension tubes; light entering the camera body, via the eyepiece, will produce high magnification, and the image size can be regulated by pushing the camera in or out. But a time exposure is essential, and I am afraid that anyone who simply holds up a camera to a telescope eye-piece and goes 'click' is doomed to disappointment.

I am well aware that these notes are hopelessly brief. Certainly there is endless scope for the astronomical photographer, whether he wants to make detailed pictures of planets of whether he is content to do no more than photograph star trails.

Appendix 30

Astronomical Societies

I have stressed the advantages of joining a society – and fortunately, there are many of them. In Britain, the leading mainly-amateur organization is the British Astronomical Association, with its headquarters at Burlington House, Piccadilly, London W1; it holds monthly meetings, publishes a regular journal, and has sections devoted to specialized branches of observation. There is no qualification for membership, other than enthusiasm, and it has an observational record second to none; in 1990 it will be a hundred years old.

The Junior Astronomical Society caters for younger enthusiasts. Ireland has its own society, with its headquarters at the Planetarium, Armagh. And all over the British Isles there are local societies, many of which have their own observatories and produce excellent work. A full list of them is given each year in the annual *Yearbook of Astronomy*.

Other countries are equally well served. There are many societies in the United States, listed in the American edition of the *Yearbook of Astronomy*; there are also many in Australia, and the Royal Astronomical Society of New Zealand is largely amateur. In South Africa there are local associations, and also the National Astronomical Society of South Africa.

Note, too, that amateurs are by no means excluded from most of the eminent professional organizations, such as Britain's Royal Astronomical Society. Indeed, the list of Past Presidents of the Royal Astronomical Society includes the names of several amateurs, though this is naturally very much the exception to the general rule.

INDEX